I0488405

About the National Science and Technology Council

The National Science and Technology Council (NSTC) was established by Executive Order on November 23, 1993. The Cabinet-level council is the principal means by which the President coordinates science and technology policies across the Federal Government. The NSTC coordinates the diverse parts of the Federal research and development enterprise. An important objective of the NSTC is the establishment of clear national goals for Federal science and technology investments in areas ranging from nanotechnology and health research to improving transportation systems and strengthening fundamental research. The Council prepares research and development strategies that are coordinated across Federal agencies to form a comprehensive investment package aimed at accomplishing multiple national goals. For more information on the NSTC, visit the NSTC website at http://www.whitehouse.gov/administration/eop/ostp/nstc.

About the Office of Science and Technology Policy

The Office of Science and Technology Policy (OSTP) was established by the National Science and Technology Policy, Organization, and Priorities Act of 1976. OSTP's responsibilities include advising the President in policy formulation and budget development on all questions in which science and technology (S&T) are important elements; articulating the President's S&T policies and programs; and fostering strong partnerships among Federal, State, and local governments, and the scientific communities in industry and academia. The Director of OSTP also serves as Assistant to the President for Science and Technology and manages the NSTC for the President. For additional information regarding OSTP, visit the OSTP website at http://www.ostp.gov.

About this document

This document is a supplement to the President's 2014 Budget Request submitted to Congress on April 10, 2013. It gives a description of the activities underway in 2012 and 2013 and planned for 2014 by the Federal Government agencies participating in the National Nanotechnology Initiative (NNI), primarily from a programmatic and budgetary perspective. It is based on the NNI Strategic Plan released in February 2011 and reports actual investments for 2012, estimated investments for 2013, and requested investments for 2014 by Program Component Area (PCA), as called for under the provisions of the 21st Century Nanotechnology Research and Development Act of 2003 (Public Law 108-153, 15 USC §7501). The report also addresses the requirement for Department of Defense reporting on its nanotechnology investments, per 10 USC §2358. Additional information regarding the NNI is available on the NNI website at www.nano.gov.

About the cover

Transmission electron micrograph (TEM) of cellulose nanocrystals (CNCs) produced from commercial eucalyptus bleached kraft pulp by strong acid hydrolysis. The CNCs have relatively uniform physical dimensions (diameter of approximately 10 nanometers and length of 50 to 100 nanometers). Because CNCs have excellent physical strength, but are light weight (density of 1.5 grams per cubic centimeter) and optically clear; they can be used in applications such as reinforced glass for composite windshields. Other potential applications include transparent films, flexible displays and electronics, reinforcing fillers for polymers, fibers and textiles, separation membranes, and biomedical implants. The CNCs were produced by researchers at the USDA Forest Service, Forest Products Laboratory, Madison, Wisconsin. The TEM of the CNCs was acquired by the Nanotechnology Characterization Laboratory of the Frederick National Laboratory for Cancer Research, funded by the National Cancer Institute.

Cover and book design

Cover design is by Kathy Tresnak of Koncept, Inc. Book design is by staff of the National Nanotechnology Coordination Office (NNCO).

Copyright information

Printed in the United States of America, 2013.

SUPPLEMENT TO THE PRESIDENT'S BUDGET FOR FISCAL YEAR 2014

THE NATIONAL NANOTECHNOLOGY INITIATIVE

Subcommittee on Nanoscale Science, Engineering, and Technology
Committee on Technology
National Science and Technology Council

May 2013

May 10, 2013

Dear Members of Congress:

I am pleased to forward this annual report on the multi-agency National Nanotechnology Initiative (NNI), in the form of the NNI Supplement to the President's Budget for Fiscal Year 2014. This document summarizes the programs and coordinated activities taking place across the many departments and agencies participating today in the NNI – an initiative that has been a leading model of Federal science and technology coordination for more than a dozen years.

The proposed NNI budget for Fiscal Year 2014 of over $1.7 billion will advance our understanding of nanoscale phenomena and our ability to engineer nanoscale devices and systems that address national priorities and global challenges. As is true across the spectrum of Federal science and technology investments, the FY 2014 NNI budget reflects difficult choices as well as the maturation and evolution of programs at some agencies. Despite current fiscal constraints, robust efforts are maintained in areas such as nanomanufacturing and major facilities, and key investments such as those in environment, health, and safety research have been increased by comparison to FY 2012 expenditures. NNI activities also intersect with and support national priorities that are not confined to the domain of nanotechnology, including advanced manufacturing, the Materials Genome Initiative (a drive to accelerate the design and deployment of advanced materials), and the recently announced BRAIN (Brain Research through Advancing Innovative Neurotechnologies) Initiative.

The NNI investment sustains vital support for fundamental, ground-breaking R&D, research infrastructure (including world-class centers, networks, and user facilities), and education and training programs that collectively constitute a major U.S. innovation enterprise. Signature initiatives within the NNI leverage programmatic collaborations among multiple agencies to place particular emphasis on jointly identified opportunities. These signature initiatives include ongoing activities in solar energy, next-generation electronics, and sustainable manufacturing as well as new activities launched in the past year in nanoinformatics and modeling and in sensors. In recognition of the special potential in these areas, the total budgeted investment in these five initiatives is slated to increase substantially in FY 2014 compared to the level of activity in FY 2012.

It is essential that the United States continue to lead the way in innovation enabled by nanotechnology and other emerging technologies. Now more than ever, the Nation's economic growth and global competitiveness depend on it. Thank you for sharing and supporting that vision.

Sincerely,

John P. Holdren

John P. Holdren
Assistant to the President for Science and Technology
Director, Office of Science and Technology Policy

TABLE OF CONTENTS

1. INTRODUCTION AND OVERVIEW

Overview of the National Nanotechnology Initiative

The National Nanotechnology Initiative (NNI), established in 2001,[1] is a U.S. Government research and development (R&D) initiative involving 27 department and agency units working together toward the shared and challenging vision of *a future in which the ability to understand and control matter at the nanoscale leads to a revolution in technology and industry that benefits society.*[2] The combined, coordinated efforts of these department and agency units have accelerated discovery, development, and deployment of nanotechnology to benefit agency missions in service of the broader national interest. The 27 department and agency units participating in the NNI are shown in Table 1; 15 of these report specific budget data for nanotechnology R&D for 2014.

The NNI is managed within the framework of the National Science and Technology Council (NSTC), the Cabinet-level council by which the President coordinates science and technology policy across the Federal Government. The Nanoscale Science, Engineering, and Technology (NSET) Subcommittee of the NSTC's Committee on Technology coordinates planning, budgeting, program implementation, and review of progress for the initiative. The NSET Subcommittee is composed of representatives from participating agencies and the Executive Office of the President. A listing of official NSET Subcommittee members is provided at the front of this report, and contact information for NSET Subcommittee participants is provided in Appendix C. The National Nanotechnology Coordination Office (NNCO) acts as the primary point of contact for information on the NNI; provides technical and administrative support to the NSET Subcommittee, including the preparation of multiagency planning, budget, and assessment documents; develops, updates, and maintains the NNI website http://nano.gov; and provides public outreach on behalf of the NNI.

What is Nanotechnology?

Nanotechnology is the understanding and control of matter at dimensions between approximately 1 and 100 nanometers, where unique phenomena enable novel applications. Encompassing nanoscale science, engineering, and technology, nanotechnology involves imaging, measuring, modeling, and manipulating matter at this length scale.

A nanometer is one-billionth of a meter. A sheet of paper is about 100,000 nanometers thick; a single gold atom is about a third of a nanometer in diameter. Dimensions between approximately 1 and 100 nanometers are known as the nanoscale. Unusual physical, chemical, and biological properties can emerge in materials at the nanoscale. These properties may differ in important ways from the properties of bulk materials and single atoms or molecules.

The NSET Subcommittee has established four working groups to support key NNI activities that benefit from focused interagency attention. These are the Global Issues in Nanotechnology (GIN) Working Group; the Nanotechnology Environmental and Health Implications (NEHI) Working Group; the Nanomanufacturing, Industry Liaison, and Innovation (NILI) Working Group; and the Nanotechnology Public Engagement and Communications (NPEC) Working Group.

---

[1] **General note:** In conformance with Office of Management and Budget style, references to years in this report are to fiscal years unless otherwise noted.

[2] *The National Nanotechnology Initiative Strategic Plan* (NSTC, Washington DC, 2011; http://nano.gov/node/581), p. 4.

Table 1: Federal Department and Agency Units[3] Participating in the NNI in 2013
15 Federal departments and agencies with budgets dedicated to nanotechnology research and development

Consumer Product Safety Commission (CPSC)
Department of Homeland Security (DHS)
Department of Commerce (DOC)
    National Institute of Standards and Technology (NIST)
Department of Defense (DOD)
Department of Energy (DOE)
Department of Transportation (DOT)
    Federal Highway Administration (FHWA)
Environmental Protection Agency (EPA)
Department of Health and Human Services (HHS)
    Food and Drug Administration (FDA)
    National Institutes of Health (NIH)
    National Institute for Occupational Safety and Health (NIOSH)
National Aeronautics and Space Administration (NASA)
National Science Foundation (NSF)
U.S. Department of Agriculture (USDA)
    Agricultural Research Service (ARS)
    Forest Service (FS)
    National Institute of Food and Agriculture (NIFA)

12 other participating departments and agencies

Department of Commerce
    Bureau of Industry and Security (BIS)
    Economic Development Administration (EDA)
    U.S. Patent and Trademark Office (USPTO)
Department of Education (DOEd)
Department of Interior (DOI)
    U.S. Geological Survey (USGS)
Department of Justice (DOJ)
Department of Labor (DOL)
    Occupational Safety and Health Administration (OSHA)
Department of State (DOS)
Department of the Treasury (DOTreas)
Director of National Intelligence (DNI)
Nuclear Regulatory Commission (NRC)
U.S. International Trade Commission (USITC)[4]

---

[3] Hereinafter referred to in this document for convenience as —agencies" or —departments and agencies"

[4] Non-voting member

The 2011 NNI Strategic Plan sets out the vision for the NNI stated above and specifies four goals aimed at achieving that overall vision:

1. Advance a world-class nanotechnology research and development program
2. Foster the transfer of new technologies into products for commercial and public benefit
3. Develop and sustain educational resources, a skilled workforce, and the supporting infrastructure and tools to advance nanotechnology
4. Support responsible development of nanotechnology

For each of the goals, the plan identifies specific objectives toward collectively achieving the NNI vision. Chapter 3 provides details about these objectives and related NNI activities.

The plan also lays out eight R&D investment categories. These Program Component Areas (PCAs) include research and development activities that contribute to one or more of the NNI goals:

1. Fundamental Nanoscale Phenomena and Processes
2. Nanomaterials
3. Nanoscale Devices and Systems
4. Instrumentation Research, Metrology, and Standards for Nanotechnology
5. Nanomanufacturing
6. Major Research Facilities and Instrumentation Acquisition
7. Environment, Health, and Safety
8. Education and Societal Dimensions

The NNI R&D investment is also guided by the 2011 NNI Environmental, Health, and Safety Research Strategy.[5] The strategy supports all four NNI goals but is most closely aligned with NNI Goal 4 and PCAs 7 and 8.

The NNI funding represents the sum of the nanotechnology-related funds allocated by each of the participating agencies (the ―NNI budget crosscut"). Each agency separately determines its budget for nanotechnology R&D in coordination with the Office of Management and Budget (OMB), the Office of Science and Technology Policy (OSTP), and Congress. The NNI agencies participating in the budget crosscut work closely with each other to create an integrated scientific program. This close communication, facilitated through the NSET Subcommittee and its working groups, has led to interagency coordination and collaboration in a variety of forms, including sharing of knowledge and expertise; joint sponsorship of solicitations and workshops; and leveraging of funding, staff, and equipment/facility resources at NNI participating agencies.

Purpose of this Report

This document provides supplemental information to the President's 2014 Budget and serves as the Annual Report on the NNI called for in the 21st Century Nanotechnology Research and Development Act (P.L. 108-153, 15 USC §7501). The report also addresses the requirement for Department of Defense reporting on its nanotechnology investments, per 10 USC §2358 (see Appendix A). In particular, the report summarizes NNI programmatic activities for 2012 and 2013, as well as those currently planned for 2014.

---

[5] *The National Nanotechnology Initiative Environmental, Health, and Safety Research Strategy* (NSTC, Washington DC, 2011; http://nano.gov/node/681).

NNI budgets for 2012–2014 are presented by agency and PCA in Chapter 2 of this report. Information on the use of the Small Business Innovation Research (SBIR) and Small Business Technology Transfer Research (STTR) program funds to support nanotechnology research and commercialization activities, also called for in P.L. 108-153, is included at the end of Chapter 2. Activities that have been undertaken and progress that has been made toward achieving the four goals set out in the NNI Strategic Plan, activities in support of the NNI Nanotechnology Signature Initiatives (NSIs), changes in the balance of investments by PCA, and highlights from external reviews of the NNI and how their recommendations are being addressed, are presented in Chapters 3–6.

2. NNI INVESTMENTS

Budget Summary

The President's 2014 Budget provides over $1.7 billion for the National Nanotechnology Initiative (NNI), a sustained investment in support of the President's priorities and innovation strategy. Cumulatively totaling almost $20 billion since the inception of the NNI in 2001 (including the 2014 request), this support reflects nanotechnology's potential to significantly improve our fundamental understanding and control of matter at the nanoscale and to translate that knowledge into solutions to critical national issues. NNI research efforts are guided by two strategic documents developed by the Nanoscale Science, Engineering, and Technology (NSET) Subcommittee of the National Science and Technology Council (NSTC), the 2011 NNI Strategic Plan (http://nano.gov/node/581) and the 2011 NNI Environmental, Health, and Safety Research Strategy (http://nano.gov/node/681). These strategic documents guide how NNI agencies address the full range of nanotechnology research and development, technology transfer and product commercialization, infrastructure and education, as well as the societal issues that accompany an emerging technology. The investments in 2012 and 2013 and those proposed for 2014 continue the emphasis on accelerating the transition from basic R&D to innovations that support national priorities.

The 2014 NNI budget supports nanoscale science, engineering, and technology R&D at 15 agencies. Agencies with the largest investments are:

- DOE (fundamental and applied research providing a basis for new and improved energy technologies)

- NSF (fundamental research and education across all disciplines of science and engineering)

- NIH (nanotechnology-based biomedical research at the intersection of life and physical sciences)

- DOD (science and engineering research advancing defense and dual-use capabilities)

- NIST (fundamental research and development of measurement and fabrication tools, analytical methodologies, and metrology for nanotechnology)

Other agencies investing in mission-related nanotechnology research are NASA, EPA, NIOSH, FDA, USDA (including ARS, NIFA, and FS), DHS, CPSC, and DOT (including FHWA).

Table 2 presents NNI investments in 2012–2014 for Federal agencies with budgets and investments for nanotechnology R&D. Tables 3–5 list the investments by agency and by Program Component Area (PCA). Table 6 supplies the NNI investments within Small Business Innovation Research (SBIR) and Small Business Technology Transfer (STTR) programs. Table 7 (in Chapter 4 of this report, page 50) shows estimated funding for the Nanotechnology Signature Initiatives (NSIs).[6]

---

[6] Note that for 2012–2014, NSI investments are not part of the formal NNI budget crosscut called for by OMB each year. Estimates figures shown in Table 7 cut across most of the PCA categories shown in Tables 3–5, and are not additional investments beyond the PCAs.

Table 2: NNI Budget, by Agency, 2012–2014 (dollars in millions)			
Agency	2012 Actual	2013 CR*	2014 Proposed
CPSC	2.0	2.0	2.0
DHS	18.7	18.7	34.9
DOC/NIST	95.4	93.3	102.1
DOD	426.1	214.2	216.9
DOE**	313.8	354.3	369.6
DOT/FHWA	1.0	0.0	2.0
EPA	17.5	17.5	17.2
HHS (total)	479.6	486.3	488.6
FDA	13.6	16.5	16.8
NIH	456.0	458.8	460.8
NIOSH	10.0	11.0	11.0
NASA	18.6	19.2	17.5
NSF	466.3	426.0	430.9
USDA (total)	18.3	18.3	20.3
ARS	2.0	2.0	2.0
FS	5.0	5.0	7.0
NIFA	11.3	11.3	11.3
TOTAL***	1857.3	1649.8	1702.0

*   The 2013 appropriations were not enacted at the time that the 2014 Request was formulated; therefore, the amounts in the 2013 column reflect the annualized level provided by the Continuing Resolution (CR, P.L. 112-175).

**   Funding levels for DOE include the combined budgets of the Office of Science, the Office of Energy Efficiency and Renewable Energy (EERE), the Office of Fossil Energy, and the Advanced Research Projects Agency for Energy (ARPA-E).

***   In Tables 2–6, totals may not add, due to rounding.

Key Points about the 2014 NNI Investments

- Overall, NNI investments remain an Administration priority and are essentially sustained for 2012 through 2014, despite the current austere budgetary environment. Top priority areas within the NNI investment portfolio such as environmental, health, and safety (EHS) research (PCA 7) and NSIs are increasing substantially over prior years, both in terms of actual investments for 2012 and in the President's Budget requests for 2014. Increases in nanomanufacturing (PCA 5) investments since 2006 have been sustained.[7]

- While fundamental research (PCA 1) remains the largest single NNI investment category ($445 million in the 2014 budget), the more applied research in nanodevices and systems (PCA 3) and in nanomanufacturing (PCA 5) are sustained at about $500 million combined for 2014, as some areas of nanotechnology mature and applications develop. The percentage of total NNI investments dedicated to PCA 5 has more than doubled on average since 2006.

---

[7] For comparisons to actual NNI funding by PCA prior to 2012, see http://nanodashboard.nano.gov.

Table 3: Actual 2012 Agency Investments by Program Component Area (dollars in millions)								
			3. Nanoscale Devices & Systems		5. Nano-manufacturing		7. Environment, Health, and Safety	8. Education & Societal Dimensions	
CPSC	0.0	0.0	0.0	0.0	0.0	0.0	2.0	0.0	2.0
DHS	0.0	0.0	0.0	0.0	9.8	0.0	8.9	0.0	18.7
DOC/NIST	21.5	7.3	17.8	16.1	9.2	16.4	7.2	0.0	95.4
DOD	172.4	49.6	141.9	0.5	42.2	18.5	1.0	0.0	426.1
DOE	100.0	87.5	12.4	11.2	0.0	102.8	0.0	0.0	313.8
DOT/FHWA	0.0	1.0	0.0	0.0	0.0	0.0	0.0	0.0	1.0
EPA	0.0	0.0	0.0	0.0	0.0	0.0	17.5	0.0	17.5
HHS (total)	55.4	118.2	213.4	16.5	5.7	20.6	46.6	3.1	479.6
FDA	0.0	0.0	0.0	0.0	0.0	0.0	13.6	0.0	13.6
NIH	55.4	118.2	213.4	16.5	5.7	20.6	23.0	3.1	456.0
NIOSH	0.0	0.0	0.0	0.0	0.0	0.0	10.0	0.0	10.0
NASA	0.0	9.2	5.8	0.0	2.6	0.0	0.0	1.0	18.6
NSF	167.6	78.8	62.6	13.1	44.4	38.8	24.2	36.8	466.3
USDA (total)	2.0	6.8	4.0	1.0	1.0	0.0	2.5	1.0	18.3
ARS	0.0	2.0	0.0	0.0	0.0	0.0	0.0	0.0	2.0
FS	1.0	2.0	0.0	1.0	1.0	0.0	0.0	0.0	5.0
NIFA	1.0	2.8	4.0	0.0	0.0	0.0	2.5	1.0	11.3
TOTAL	518.8	358.5	457.9	58.3	114.9	197.1	109.9	41.9	1857.3

- The NNI agencies continued to maintain the NNI research infrastructure at the cutting edge of scientific capability through ongoing upgrades to facilities and instrumentation, including user facilities available to industry and academia. NNI investments in instrumentation research, metrology, and standards (PCA 4) and in major research facilities and instrumentation acquisition (PCA 6) have averaged around $250 million combined during the 2012–2014 period. This sustained investment also reflects the importance of metrology and standards to commercialization of nanotechnology-enabled innovations.

- The NNI continues to maintain a robust and growing investment in EHS research related to nanotechnology (PCA 7), with over $120 million requested for 2014, an increase of about 10% over 2012 actual spending, and more than 37% over 2011 spending. The NNI investment in PCA 7 now represents over 7% of the total NNI budget for 2014. This increase is guided by the NNI EHS Research Strategy that was released in October 2011. Cumulatively the NNI agencies have allocated more than $750 million to EHS research since 2006, including the 2014 President's Budget Request.

	1. Fundamental Phenomena & Processes		3. Nanoscale Devices & Systems	4. Instrument Research, Metrology, & Standards	5. Nano-manufacturing	6. Major Research Facilities & Instr. Acquisition	7. Environment, Health, and Safety	8. Education & Societal Dimensions	
Table 4: 2013 CR Agency Investments by Program Component Area (dollars in millions)									
CPSC	0.0	0.0	0.0	0.0	0.0	0.0	2.0	0.0	2.0
DHS	0.0	0.0	0.0	0.0	9.8	0.0	8.9	0.0	18.7
DOC/NIST	20.4	7.2	17.8	14.7	8.9	16.4	7.9	0.0	93.3
DOD	98.1	41.7	46.1	0.4	15.1	12.1	0.7	0.0	214.2
DOE	105.6	105.2	12.7	16.7	3.0	110.9	0.0	0.2	354.3
DOT/FHWA	0.0	0.0	0.0	0.0	0.0	0.0	0.0	0.0	0.0
EPA	0.0	0.0	0.0	0.0	0.0	0.0	17.5	0.0	17.5
HHS (total)	55.8	118.9	214.7	16.6	5.7	20.7	50.7	3.2	486.3
FDA	0.0	0.0	0.0	0.0	0.0	0.0	16.5	0.0	16.5
NIH	55.8	118.9	214.7	16.6	5.7	20.7	23.2	3.2	458.8
NIOSH	0.0	0.0	0.0	0.0	0.0	0.0	11.0	0.0	11.0
NASA	0.0	8.5	7.0	0.0	2.7	0.0	0.0	1.0	19.2
NSF	146.3	78.8	52.4	12.1	47.8	28.5	30.0	30.1	426.0
USDA (total)	2.0	6.8	4.0	1.0	1.0	0.0	2.5	1.0	18.3
ARS	0.0	2.0	0.0	0.0	0.0	0.0	0.0	0.0	2.0
FS	1.0	2.0	0.0	1.0	1.0	0.0	0.0	0.0	5.0
NIFA	1.0	2.8	4.0	0.0	0.0	0.0	2.5	1.0	11.3
TOTAL	428.2	367.1	354.7	61.5	93.9	188.6	120.2	35.5	1649.8

- The NSIs are a very high priority for the NNI, as indicated by the steady increases in overall NSI investments, from $294 million in 2012 to a requested $343 million in 2014 (see Table 7 in Chapter 4 of this report, page 50). The three NSI topics initiated in 2011 (on Solar Energy, Nanomanufacturing, and Nanoelectronics) are maturing with sustained investments (at $102 million, $60 million, and $80 million, respectively, in the 2014 Budget). Additional NSIs covering R&D on Nanosensors and on Nanotechnology Knowledge Infrastructure (NKI) were initiated in calendar year 2012, with funding requests of $78 million and $23 million in the President's 2014 Budget. (See Chapter 4 of this report for details.)

- As in previous years, investment estimates reported by some agencies fluctuate from year to year. This variation is associated with several factors: (1) At most mission agencies participating in the NNI, nanotechnology is not a single programmed investment but is associated with current research needs and opportunities. Nanotechnology must compete for resources with other technologies that contribute to the agencies' missions. Nanotechnology projects have proven to be highly competitive, and generally, actual investments have exceeded the estimates, although this is not guaranteed. (2) Specific nanotechnology projects often have milestones that need to be met for continuation, so estimates cannot firmly include these resources. (3) Occasionally a substantial new thrust in nanotechnology will enhance funding in a year without significant future funding (for example, one-time expenses for purchase of instrumentation). Hence budget estimate numbers for those agencies are sometimes substantially lower than subsequently reported actual spending.

	1. Fundamental Phenomena & Processes		3. Nanoscale Devices & Systems	4. Instrument Research, Metrology, & Standards	5. Nano-manufacturing	6. Major Research Facilities & Instr. Acquisition	7. Environment, Health, and Safety	8. Education & Societal Dimensions	
Table 5: Proposed 2014 Agency Investments by Program Component Area (dollars in millions)									
CPSC	0.0	0.0	0.0	0.0	0.0	0.0	2.0	0.0	2.0
DHS	0.0	0.0	21.4	0.0	5.6	0.0	7.9	0.0	34.9
DOC/NIST	20.4	7.2	17.8	14.7	14.7	16.4	10.9	0.0	102.1
DOD	89.9	37.2	68.7	0.3	20.0	0.1	0.7	0.0	216.9
DOE	126.4	107.0	14.2	11.8	0.0	110.0	0.0	0.2	369.6
DOT/FHWA	0.0	2.0	0.0	0.0	0.0	0.0	0.0	0.0	2.0
EPA	0.0	0.0	0.0	0.0	0.0	0.0	17.2	0.0	17.2
HHS (total)	56.3	119.3	215.5	16.7	5.8	20.8	51.0	3.2	488.6
FDA	0.0	0.0	0.0	0.0	0.0	0.0	16.8	0.0	16.8
NIH	56.3	119.3	215.5	16.7	5.8	20.8	23.2	3.2	460.8
NIOSH	0.0	0.0	0.0	0.0	0.0	0.0	11.0	0.0	11.0
NASA	0.0	7.7	7.0	0.0	2.8	0.0	0.0	0.0	17.5
NSF	148.8	80.7	51.1	12.0	49.4	28.7	29.0	31.3	430.9
USDA (total)	3.0	6.8	4.0	1.0	2.0	0.0	2.5	1.0	20.3
ARS	0.0	2.0	0.0	0.0	0.0	0.0	0.0	0.0	2.0
FS	2.0	2.0	0.0	1.0	2.0	0.0	0.0	0.0	7.0
NIFA	1.0	2.8	4.0	0.0	0.0	0.0	2.5	1.0	11.3
TOTAL	**444.7**	**367.9**	**399.6**	**56.5**	**100.3**	**176.0**	**121.1**	**35.7**	**1702.0**

- DOD funding estimates for 2014, which are approximately 50% lower than 2012 actuals, reflect the maturation and completion of projects, especially at DARPA, and broadly reduced estimates associated with budget uncertainties and ongoing competitions, which cannot be counted until awards are made.

- Many of the NNI participating agencies and programs are also actively participating in and contributing to complementary and synergistic Administration R&D priorities, including the Networking and Information Technology (NITRD) initiative, the Global Change Research Program (GCRP), the Materials Genome Initiative (MGI), Advanced Manufacturing, and the new Brain Research through Advancing Innovative Neurotechnologies (BRAIN) initiative. Nanoscale science and technology is related to, enabling of, or enabled by, each of these other priority topics in various ways, and NNI investments may overlap with investments reported for each of them. One example is nanomanufacturing research, which is reported by agencies both under the NNI (PCA 5) and under Advanced Manufacturing. Another example is the important contribution of high-end computing research, reported under the NITRD program, to modeling and simulation of nanomaterials, reported under the NNI crosscut (and to some extent under the Nanotechnology Knowledge Infrastructure NSI). Similarly, nanoelectronics research (reported under NNI PCA 3 and to some extent under the Nanoelectronics NSI) is enabling continued progress in high-end computing, reported under NITRD. A final example would be the close relationships and synergies between nanomaterials modeling and simulation R&D conducted under the NKI NSI and R&D coordinated under the MGI. NNI agency representatives and NNCO staff members are actively engaged in coordinating across all of these complementary activities to take advantage of these synergies.

- The Department of Homeland Security is reporting a significant new funding request under PCA 3 (Nanoscale Devices and Systems) for 2014.

- Investments in SBIR and STTR funding by the participating agencies, reported outside of the formal NNI funding crosscut and the budget tables shown above, play a critical role in transitioning nanotechnology innovations into products for commercial and public benefit (NNI Goal 2), as discussed below.

Utilization of SBIR and STTR Programs to Advance Nanotechnology

As called for by the 21st Century Nanotechnology Research and Development Act, this report includes information on use of the Small Business Innovation Research and Small Business Technology Transfer programs in support of nanotechnology development. Five NNI agencies—DOD, NSF, NIH, DOE, and NASA—have both SBIR and STTR programs. In addition, NIOSH, EPA, USDA, and NIST have SBIR programs. Table 6 shows agency funding for SBIR and STTR awards for nanotechnology R&D from 2007 through 2011 (the latest year for which data are available).

Some NNI agencies (e.g., NSF and NIH) have nanotechnology-specific topics in their planned SBIR and STTR solicitations. The NSF SBIR program has an ongoing nanotechnology topic with subtopics for nanomaterials, nanomanufacturing, nanoelectronics and active nanostructures, nanotechnology for biological and medical applications, and instrumentation for nanotechnology. Some agencies have had additional topical or applications-oriented solicitations for which many awardees have proposed nanotechnology-based innovations. SBIR/STTR data for 2004–2011 indicates that the NNI agencies have funded over $700 million of nanotechnology-related SBIR and STTR awards. Data from 2007–2011 are shown in Table 6, and a complete listing by year (including data from 2004, 2005, and 2006) can be found at a download link in the Overview section of http://nanodashboard.nano.gov.

Table 6: 2007–2011 Agency SBIR and STTR Awards (dollars in millions)														
	2007			**2008**			**2009**			**2010**			**2011**		
	SBIR	**STTR**	**Total**	**SBIR**	**STTR**	**Total**	**SBIR**	**STTR**	**Total**	**SBIR**	**STTR**	**Total**	**SBIR**	**STTR**	**Total**
DOD	8.4	4.2	12.6	19.8	2.3	22.1	19.3	3.9	23.2	32.2	10.4	42.6	16.2	5.6	21.8
NSF	13.4	3.8	17.2	10.5	7.5	18.0	13.2	7.0	20.2	25.2	3.9	29.2	13.2	1.3	14.5
DHHS(NIH)	18.4	1.1	19.5	29.3	1.8	31.1	23.0	3.7	26.7	21.7	1.6	23.2	22.6	1.5	24.1
DHHS(NIOSH)	0.1	0.0	0.1	0.4	0.0	0.4	0.7	0.0	0.7	0.0	0.0	0.0	0.2	0.0	0.2
DOE[8]	17.4	0.8	18.2	13.8	2.7	16.5	31.9	5.7	37.6	19.4	4.4	23.8	4.7	1.1	5.8
NASA	11.7	1.5	13.2	6.2	0.8	7.0	15.0	1.7	16.7	11.6	2.1	13.7	12.1	3.7	15.8
EPA	0.5	0.0	0.5	0.7	0.0	0.7	0.7	0.0	0.7	0.4	0.0	0.4	0.9	0.0	0.9
USDA[9]	1.1	0.0	1.1	0.9	0.0	0.9	0.5	0.0	0.5	0.0	0.0	0.0	1.0	0.0	1.0
DOC (NIST)	0.3	0.0	0.3	0.4	0.0	0.4	0.4	0.0	0.4	0.2	0.0	0.2	0.3	0.0	0.3
TOTAL	**71.3**	**11.4**	**82.7**	**81.9**	**15.1**	**97.0**	**104.6**	**22.0**	**126.6**	**110.6**	**22.4**	**133.0**	**71.2**	**13.2**	**84.4**

---

[8] A significant portion of SBIR/STTR funding reported by DOE for 2009 and 2010 is from the American Recovery and Reinvestment Act (ARRA) of 2009.

[9] SBIR/STTR award levels for USDA include FS.

3. PROGRESS TOWARDS ACHIEVING NNI GOALS, OBJECTIVES, AND PRIORITIES

Activities Relating to the Four NNI Goals and Sixteen Objectives

As called for in the 21st Century Nanotechnology Research and Development Act, this NNI Supplement to the President's Budget provides an analysis of the progress made toward achieving the goals established for the National Nanotechnology Initiative. The NNI's activities for 2012 and 2013 and plans for 2014 are reported here in terms of how they promote progress toward the four NNI goals and sixteen objectives set out in the 2011 NNI Strategic Plan (Figure 1). It is important to note that many agency programs and activities, although listed below under their primary NNI goals and objectives, address components of multiple goals simultaneously. Therefore an integrated perspective across the four goals is needed to understand progress towards achieving the NNI vision.

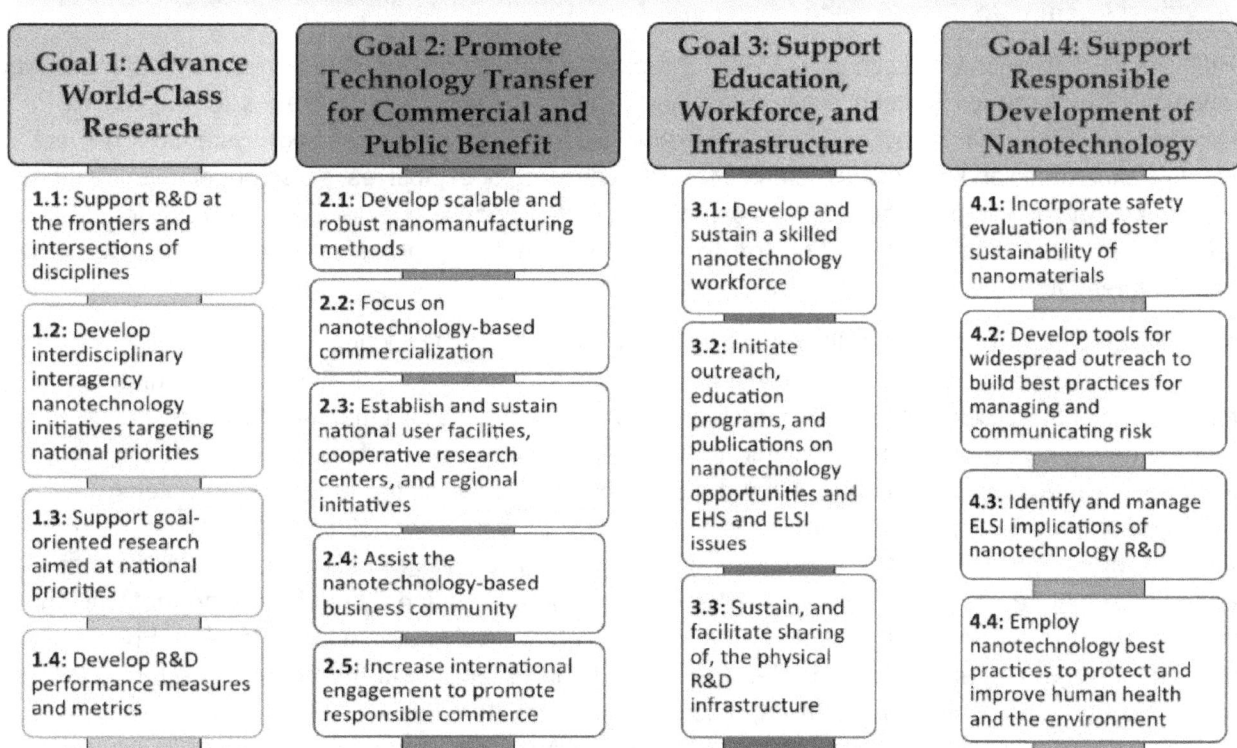

Research and Development Leading to a Revolution in Technology and Industry

Goal 1: Advance World-Class Research	Goal 2: Promote Technology Transfer for Commercial and Public Benefit	Goal 3: Support Education, Workforce, and Infrastructure	Goal 4: Support Responsible Development of Nanotechnology
1.1: Support R&D at the frontiers and intersections of disciplines	2.1: Develop scalable and robust nanomanufacturing methods	3.1: Develop and sustain a skilled nanotechnology workforce	4.1: Incorporate safety evaluation and foster sustainability of nanomaterials
1.2: Develop interdisciplinary interagency nanotechnology initiatives targeting national priorities	2.2: Focus on nanotechnology-based commercialization	3.2: Initiate outreach, education programs, and publications on nanotechnology opportunities and EHS and ELSI issues	4.2: Develop tools for widespread outreach to build best practices for managing and communicating risk
	2.3: Establish and sustain national user facilities, cooperative research centers, and regional initiatives		4.3: Identify and manage ELSI implications of nanotechnology R&D
1.3: Support goal-oriented research aimed at national priorities	2.4: Assist the nanotechnology-based business community	3.3: Sustain, and facilitate sharing of, the physical R&D infrastructure	4.4: Employ nanotechnology best practices to protect and improve human health and the environment
1.4: Develop R&D performance measures and metrics	2.5: Increase international engagement to promote responsible commerce		

Figure 1. A summary of NNI goals and objectives from the 2011 NNI Strategic Plan.[10]

Goal- and objective-related activities are reported below in terms of two categories of activities: (1) individual agency activities and (2) coordinated activities of NNI agencies with other agencies and with groups external to the NNI, including international activities. *The activities described below in terms of these*

---

[10] The NNI goals and objectives summarized in Figure 1 are adapted from the 2011 *National Nanotechnology Initiative Strategic Plan,* http://nano.gov/node/581 or http://nano.gov/about-nni/what/vision-goals.

two categories are only selected highlights of current and planned work of the NNI member agencies and are not an all-inclusive description of ongoing NNI activities.

Goal 1: Advance a world-class nanotechnology research and development program

This NNI goal seeks to expand the boundaries of scientific knowledge and to develop practical new technologies through comprehensive and focused nanotechnology research and development at the frontiers and intersections of scientific disciplines. The member agencies have sustained a strategic investment in fundamental nanotechnology R&D, an essential component for the development and successful exploitation of any emerging technology. The expanded investment by this Administration in nanotechnology will build on the foundation established over the past twelve years and set the stage for transitioning basic research discoveries into technologies and products for the benefit of our society and the Nation's economy.

Objective 1.1: Continue to support R&D at the frontiers and intersections of scientific disciplines in the form of intramural and extramural programs targeting single investigators, multi-investigator and multidisciplinary research teams, and centers for focused research.

Individual Agency Contributions to Objective 1.1

DHS: The DHS Science and Technology (S&T) Directorate plans to continue the development of micro- and nanoscale structures for electron emitters in high-performance mass spectrometers for explosives trace detection systems and X-ray sources in next-generation baggage screening systems. The use of nanofibrous organic and polymeric materials for vapor sensing of explosives is also under investigation for potential applications in handheld explosives trace detection systems. The current focus is on development and commercialization of these component technologies into original equipment manufacturer (OEM) systems for DHS deployment.

DOD: Nanotechnology research at the Army Engineer Research and Development Center (ERDC) is focused on (1) the controlled design and assembly of nanomaterials for military applications; and (2) understanding the releases of nanoscale particles over the technology life cycle to assist materiel developers in balancing performance improvement with associated environmental risks. These efforts have resulted in tools and knowledge to develop and acquire Army materiel that pose the minimum hazard to soldiers and the environment; advancements in the physical properties of carbon-nanotube-based fibers; and novel methods for nanomaterial analysis. ERDC is exploring characteristics of natural nanomaterials that have exceptional mechanical properties in order to develop the foundational understanding that will lead to advances in blast and ballistic protection through engineered material models. The ERDC is also investigating a quantitative means to determine the environmental and human health effects resulting from exposure to explosives; propellants; smokes; and products containing nanomaterials and new and emerging compounds and materials produced or used in Army industrial, field, and battlefield operations or disposed of through past activities. Methods are being explored for the design of nanomaterials and other new and emerging materials such that adverse effects on human health or the environment are minimized in their designed state and when they enter the environment where they may break down (also relevant to Objective 4.4).

The Navy maintains an active collection of efforts to identify, understand, and develop nanomaterial technologies, taking advantage of the unique properties they demonstrate. A new Office of Naval Research (ONR) program was initiated to study tailoring of atomic-scale interphase complexions for mechanism-informed material design. This Multidisciplinary University Research Initiative (MURI) project is exploring nanoscale interphase and grain boundaries to tailor desirable materials properties. To meet these ambitious goals, a highly integrated experimental and computational program has been implemented that brings together the leading researchers in the field. Multiscale computations combining density functional theory, molecular dynamics, phase field, Potts model, and thermodynamic models are being correlated with multiscale materials characterization.

Air Force research in nanomaterials and nanodevices continues to offer new defense capability and application potential by improving response, efficiency, and properties in materials systems. By identifying materials properties in defense technologies with specific Air Force relevance, research programs have been targeted to investigate potentially revolutionary advances in performance by novel means such as through graphene metamaterials, novel active ceramic materials for laser sources, and high-performance nanotube fibers. Notable advances, by collaborative industrial–governmental–academic research teams that combine self-assembled nanosystems and optical metamaterials, have received significant attention at national and international conferences. Academic partnerships in nanomagnetics have led to significant progress toward nanoparticle magnetic inks that will enable additive magnetic patterning on structural components for electromagnetic applications, including conformal antennas.

Microelectronics innovation has the potential to carry complementary metal-oxide semiconductor (CMOS) to its physical limits and to provide solutions to the difficulties inherent in applying integrated circuit technology to complete systems. The Focus Center Research Program (FCRP) is a collaborative effort between the Defense Advanced Research Projects Agency (DARPA) and leading companies from the semiconductor and defense industries supporting basic research at U.S. universities. Under FCRP, research focuses on leveraging complex systems integration to provide new capabilities, combined with identifying new materials and device physics to accelerate progress for new mainstream technologies beyond today's integrated circuits. Recent key accomplishments under this program include the realization and commercialization of power-efficient, three-dimensional, multigate FinFET transistors, which Intel Corporation calls the "most radical shift in semiconductor technology in 50 years"; the demonstration of the first silicon-compatible germanium quantum well waveguide modulator with a 7-gigabit/second transfer data rate for photonic IC interconnects; and the brain–machine control of external devices by monkeys by formation of dedicated motor memory neural circuits (also relevant to Objective 1.2).

DARPA's Quantum-Assisted Sensing and Readout (QuASAR) program is developing nanoscale magnetic resonance imaging (MRI) probes for biological systems. Such devices could resolve individual nuclear spins at the nanoscale, enabling the detailed 3D mapping of biological molecules with elemental specificity in their native environment. This information could streamline assessment of inhibitor drugs against viruses, whether naturally occurring or engineered as biological weapons. The nanoscale magnetometers developed under QuASAR are based on nitrogen-vacancy (NV) impurities in diamond. Like atoms, NV impurities have exceptionally long coherence times but are localized within the rigid lattice of a diamond crystal, making them ideal for high-resolution MRI and compatible with biological systems. Recently, two QuASAR research teams demonstrated the highest-resolution MRI to date—five cubic nanometers—more than 1000 times smaller in volume than conventional MRI. This work was published recently in *Science* magazine (February 2013).

The Defense Threat Reduction Agency (DTRA) Basic and Applied Sciences program supports investigations of nanoscience that may lead to the development of nanotechnologies applicable to a number of counter-weapons of mass destruction (WMD) mission needs, including those related to securing and monitoring WMDs, systems and personnel protection, production of energetic and non-energetic materials for targeted effects, sensing of radiological and nuclear materials, and event monitoring/post-event forensics.

Programmatic interests are supported through current investment in exploratory research efforts in areas such as (1) engineered nanomaterials for unattended sensing, (2) nanostructures as novel radiation indicators, (3) nanocomposites as radiation-sensitive materials, (4) nanoscience as a building block for novel metamaterials, and (5) nanoparticle scintillators for radiation detection.

The Chemical and Biological Defense Program (CBDP) exploratory single-investigator and multi-investigator investments have been made in the programmatic areas of (1) advanced nanotechnology-enabled sensing and diagnostics platforms, including those that enlist new understanding of common biological mechanisms of threat action as well as novel fluidics and plasmonic approaches; (2) nanostructured materials for control of micro- and nanoscale separation and transport; and (3) understanding and modulation of biotic/abiotic interfaces. These areas have been identified as strategically relevant to development of future nanotechnology-enabled defense needs such as threat sensing, point-of-care diagnostics, and threat countermeasures.

DOE: The DOE Office of Science (DOE/SC) supports a broad and diverse investment in nanoscience research and development, primarily through grants to individual researchers at universities, funding for research groups at DOE laboratories, and support of interdisciplinary efforts such as the Energy Frontier Research Centers (EFRCs). DOE-SC continues to support the Joint Center for Artificial Photosynthesis (JCAP), an Energy Innovation Hub aimed at developing revolutionary methods to generate fuels directly from sunlight. It is estimated that approximately 20% of the activities in this hub involve nanotechnology. In 2013, DOE-SC initiated the Joint Center for Energy Storage Research (JCESR), an Energy Innovation Hub focused on batteries and energy storage, and it is anticipated that a significant fraction of that effort will also involve nanotechnology. By fostering unique, cross-disciplinary collaborations, these hubs will help advance highly promising areas of energy science and engineering from the early stage of research to the point where the technology can be handed off to the private sector.

In 2012, the Solar Energy Technology Program (SETP) in the DOE Office of Energy Efficiency and Renewable Energy (EERE) supported a variety of efforts related to nanotechnology research and development at the intersection of scientific disciplines. For example, the BRIDGE Funding Opportunity Announcement (FOA) was released to support collaborative efforts between scientists with photovoltaic experience and those with expertise utilizing the national scientific user facilities. The BRIDGE FOA was executed as a collaborative effort between DOE-SETP and DOE-SC. A variety of projects funded under this FOA are related to nanotechnology, such as those focused on grain boundaries and nanoscale arrays of photoactive materials.

In 2012, the Advanced Research Projects Agency–Energy (ARPA-E) created multiple new programs that involve R&D in nanotechnology. ARPA-E created the Methane Opportunities for Vehicular Energy (MOVE) program, which focuses on developing innovative, low-cost natural gas storage technologies and methods that will help enable the widespread adoption of natural gas vehicles. ARPA-E's open funding opportunity program focuses on a wide array of energy technologies, including organic photovoltaics, building efficiency, window coatings, and energy storage. In 2013, ARPA-E aims to focus on the development of disruptive technologies that can significantly reduce the energy consumption needed to extract light metals (aluminum, titanium, and magnesium) while simultaneously enabling high-grade heat recovery for power generation. ARPA-E may also invest in the field of tribology and in reducing surface drag in a variety of settings.

In 2014, ARPA-E plans to focus on a variety of new energy technologies that will rely on nanostructured materials, nanoscale manufacturing, and integration into systems. Some possible technologies include (a) biogenic sources of natural gas and low-cost, high-density storage of natural gas for transportation; (b) non-Fischer-Tropsch-based conversion of natural gas to infrastructure-compatible liquid transportation fuels; (c) low-cost and high-performance nanostructured membranes for desalination and osmotic power generation; (d) integration and packaging of low-cost, lightweight energy storage systems for electric vehicles; and (e) bioreactor systems for scaling nonphotosynthetic biofuels.

DOT/FHWA: The Federal Highway Administration (FHWA) is investigating nanoscale approaches for improving durability and mitigating degradation of structural materials. Focus areas include detection and mitigation of corrosion as well as improvement of cementitious materials by developing characterization and modification techniques for enhancing interface behavior and expanding the fundamental understanding of cement hydration kinetics. These efforts, combined with better multiscale modeling capabilities to predict material response under complex loads, aim to significantly improve the life-cycle performance of transportation infrastructure.

IC/DNI: The National Reconnaissance Office (NRO) continues to work on enhancing the properties of carbon nanotubes (CNTs), specifically in three critical areas:

- Growth of longer CNTs with a goal of 7 mm: micrographs show we have achieved 7.7 to 8.2 mm CNTs in the laboratory.

- Better alignment of CNTs during growth: initial alignment has been proven, but much work is needed in this area.

- Increased density of CNTs in sheets and yarns: dramatic improvements of 6–10-fold have been made; these products will undergo commercialization in calendar year 2013. These improvements will lower the cost of these products.

Ongoing research into decreasing direct current (DC) resistivity for 48-volt DC data center operations is within 1/15 of copper wire performance and making great strides; the goal is to save 10 to 15% energy consumption in every U.S. data center. Nonvolatile carbon-nanotube-based memory has now been embraced by IMEC in Europe, a rival to SEMATECH International and several U.S. memory/storage companies. The goal is to eliminate all disk drives and flash memory buffers from data centers for huge cost savings and for preventing technology refresh costs and the periodic transfer of data, while speeding up processing 100-fold.

Academic research continues to tackle the challenge of finding a nano-reinforcement-compatible resin that is optimized for CNT composites for superior strength and enhanced acoustic dampening. Technical discussions for applying CNT composites to U.S. space launch vehicles has been initiated with NASA and NRO, with the possibility of a demonstration on a NASA sounding rocket.

The NRO continues to make progress in developing superior-strength CNT bulk materials with improved electrical and thermal conductivity.

NASA: NASA supports a broad portfolio of research in nanostructured materials and nanotechnology-based sensors and devices. Nanotechnology research is supported by the Aeronautics, Human Exploration and Operations, and Space Technology Mission Directorates. Research in nanostructured materials is focused primarily on the development of lightweight, multifunctional materials for use in aircraft and spacecraft. Technical accomplishments for 2012 include the following:

- A new start in the development of CNT-based high-strength reinforcements as replacements for carbon fiber in composites was initiated under the Space Technology Program. This activity, focused on improving the tensile properties of dry-spun carbon nanotube sheets, tapes, and yarns, involves work at five NASA centers, industry, and academia, and is a collaboration between NASA, the Air Force Office of Scientific Research (AFOSR), and other DOD organizations. As a part of this effort, NASA co-sponsored a workshop with the Air Force Research Lab (AFRL) and AFOSR on ─Nanotube Assemblages for Structures" at the Georgia Institute of Technology on April 17–18, 2012. This workshop focused on identifying the technical barriers to developing carbon-nanotube-based high-strength materials.

Follow-on workshops are planned, including one on modeling and simulation of carbon nanotube growth and mechanical behavior planned for the spring of 2013 (also applicable to Objective 1.2).

- Research in nanoporous structural materials has led to the development of polyimide aerogels for use as thermal insulation and as substrates for antennas for aircraft and spacecraft. A team from NASA and the Ohio Aerospace Institute received an R&D-100 award for this technology in 2012. NASA is currently working with DOD and industry to transition this technology to other applications.

Nanotechnology-based sensor R&D is focused on the development of minimally invasive sensors for vehicle structural health monitoring and of sensors for the detection of chemical and biological species for use in planetary exploration, astronaut health management, and propulsion systems. Significant activities and accomplishments in 2012 include:

- Initiation in 2012 of development of nanotechnology-based sensors for the detection of strain and/or early onset of damage in composites. This effort involves work at two NASA centers, the Jet Propulsion Lab, and academia and is focused primarily on the development of CNT- and graphene-based devices.

- Development of new sensors capable of room-temperature detection of methane gas. These sensors utilize porous tin oxide nanorods, synthesized from multi-walled carbon nanotube templates, as the electrode material. The work was recently published in *Nanotechnology* (October 2012).

NIH: The NIH's nanoscience and nanotechnology initiatives and programs are funded through investigator-initiated grants as well as institute-initiated funding opportunities, with research funding focused on the government priorities and/or scientific gaps that are described in the PCA section of this report (Chapter 5). Research on emerging technologies, projected for 2014, will focus on the translation of nanoscience discoveries into new biomedical technologies. In addition to research on using nanotechnology to deliver new therapies, several of NIH's institutes are focused on advancing nanotechnology-based techniques, including clinical lab tests and assays, and tools and devices, for early disease diagnosis; and *in vivo* imaging techniques. Medical imaging is not only visualizing specific anatomic structures but also the molecular events that precede advanced symptoms. This search for biomarkers that are extremely sensitive and highly specific to detect trace molecules that signal the presence of disease or infection before a patient has become severely ill remains an active area of research for NIH. Among current NIH initiatives, the most recent ones were released in 2012 with a promising outlook for the future:

- In 2012, the National Cancer Institute (NCI) issued two new RFAs (requests for applications) to stimulate research in significant unanswered questions about cancer, backed by $24 million in set-aside funding. This "Provocative Questions" (PQs) initiative (RFA-11-011, RFA-11-012) issued 24 research questions, many of which were relevant to the biomedical nanotechnology community. Six of these PQs address PCA 3 and one addresses PCA 7. Of the 56 funded grants, eight use nanotechnology and will receive about $3.3 million in 2014. The success and interest in this initiative has led to a reissuance of the RFA with set-aside funding of $28–$40 million. Of this new set of 24 PQs, eight are relevant to nanotechnology (six to PCA 3 and two to PCA 7), and NCI expects to support additional projects relevant to NNI PCAs.

- In 2012, the NIH Roadmap for Medical Research program awarded $7.3 million to fund grants sponsored by the National Institute of Neurological Disorders and Stroke (NINDS) for neurodegenerative diseases and repair pertaining to PCAs 1–3.

These new initiatives, as well as the support for ongoing major investment programs such as the NIH-wide Nanomedicine Development Centers Common Fund Program, the NCI Alliance for Nanotechnology in Cancer, and the National Heart, Lung, and Blood Institute (NHLBI) Programs of Excellence in Nanotechnology, have been very successful in advancing the science, understanding, and applications of nanomaterials.

The NHLBI continues to support the Programs of Excellence in Nanotechnology (PENs). The PENs were renewed in 2010 and encompass 18 institutions nationwide as part of four contracts. Proposals for the NHLBI PEN program were required to include translational projects leading to investigational new drug (IND) applications.

The NCI Alliance for Nanotechnology in Cancer program, which supports 34 grants for multi-institute transdisciplinary centers, is rapidly transforming advances in nanotechnology discoveries into cancer-relevant applications with clinical utility. Advancements for 2013–2014 will include:

- Development of triple-modality nanoparticles to monitor brain surgery and combining magnetic resonance, photoacoustic, and Raman imaging.
- Development of nanoparticle delivery vehicles with very high drug loading, exceeding 50%.
- Microfluidic systems for precise synthesis of nanoparticles containing multiple drugs to be used in combination therapies.

The eight Nanomedicine Development Centers supported by the NIH Common Fund Program are developing new approaches and technologies for potential therapeutic regimens. Research supported by these centers focuses on improving our understanding of how the biological machinery inside living cells is built and operates at the nanoscale, and using this information to reengineer these structures toward therapeutic aims. The centers are developing new technologies that could be applied to treating diseases and/or to leveraging new knowledge to focus work directly on translational studies to treat disease or repair damaged tissue.

NIST: NIST is developing the quantitative atom probe, advanced electron and scanning probe microscopy, and optical nanocalorimetry measurement instrumentation needed to advance technologies ranging from semiconductors and data storage to biotechnology and optoelectronics. Measurement methods and instrumentation are being developed for nanoparticle metrology and characterization, including separation, particle tracking, surface composition, and methods to observe structural changes during chemical reactions. NIST is developing new methods for nanometer-scale dimensional metrology and imaging. It has achieved unprecedented quality and repeatability in scanning electron microscopy; developed high-throughput optical techniques with sub-nanometer sensitivity for critical defect detection and dimensional control of nanostructures; developed small-angle X-ray scattering for nondestructive, 3D profiling of nanoscale patterns; and used atomic-scale patterning methods to create a new, comprehensive approach to dimensional metrology based on using the crystal lattice as a fundamental "ruler" for surface metrology.

NIST is supporting high-risk, leading-edge nanotechnology research that anticipates the future measurement and standards needs of industry and science through the Innovations in Measurement Science (IMS) Program. Multidisciplinary IMS research teams are bridging measurement length scales to advance graphene device technologies and are developing the measurement methods needed to characterize in three dimensions the essential nanoscale processes in photovoltaic materials and devices.

NSF: NSF supports nanoscale science and engineering throughout all of its research and education directorates as a means to advance discovery and innovation and integrate various fields of research. NNI enables increased interdisciplinarity at the atomic and molecular levels for about 5,000 active awards with content in nanoscale science and engineering.

Converging science and engineering at the nanoscale work supports the convergence of nanotechnology with information technology, modern biology, and social sciences, potentially reinvigorating discoveries and innovation in almost all areas of the economy. Examples are programs investigating the nano–biology interface, the nano–information interface, and nanotechnology-related social sciences.

Multiscale, multiphenomena theory, modeling, and simulation at the nanoscale work supports theory, modeling, large-scale computer simulation and new design tools, and infrastructure in order to understand, control, and accelerate development in new nanoscale regimes and systems. An example is the recompetition and award of the Network for Computational Nanotechnology in 2012–2013.

USDA/FS: Cellulose nanomaterials have emerged as a renewable material that will enable our Nation's future to be a green and sustainable society. The Forest Service (FS) Forest Products Laboratory (FPL), partnering with an Agenda 2020-led public–private partnership (PPP), continues to advance multidisciplinary cellulose nanomaterials research in aircraft materials, predictive modeling, sensors, and composites. The PPP was formally funded in 2011, and research in the PPP is progressing toward the common vision established by FS and industrial partners. Other nanotechnology research in the FS FPL has shown that nanoscale zinc oxide is more effective than conventional zinc oxide in protecting wood against termites. In 2012, a member of the FS FPL won the Presidential Early Career Award for Scientists and Engineers for his work on developing a methodology to understand material properties at the nanoscale.

USDA/NIFA: The National Institute of Food and Agriculture (NIFA) nanotechnology efforts continue to advance the frontiers of interdisciplinary nanoscale science, engineering, and technology R&D to address a broad range of grand societal challenges and opportunities facing agriculture and food systems through its various extramural research funding mechanisms. Its nanotechnology program supports innovative ideas to develop nanotechnology-enabled solutions for global food security through improving productivity and quality; adaptation and mitigation of agricultural production systems to climate changes; improvement of the nutritional quality of food and feed; early detection and effective intervention technologies for ensuring food safety and biosecurity; and development of new biology-based products and energy solutions (also applies to Objective 1.3). NIFA's flagship competitive grants program, the Agriculture and Food Research Initiative (AFRI) Foundational Program, has combined the funding resources from 2012 and 2013 and issued one combined RFA whose nanotechnology program is constructed within the broad program scope stated above.

Coordinated Activities with Other Agencies and Institutions Contributing to Objective 1.1

DOD, NNI agencies, universities, industry: Agencies and institutions that perform nanotechnology and nanomaterial collaborative research with the ERDC are Rice University for material synthesis and atomistic modeling; the University of Southern California for atomistic and *ab initio* modeling; Oak Ridge National Laboratory for material diagnostics; the University of California Center for Environmental Implications of Nanomaterials (CEIN) for modeling, fate, and toxicity of nanoparticles (funded by NSF and EPA); the Duke University Center for Environmental Implications of NanoTechnology (CEINT) for particle analysis and toxicity (funded by NSF and EPA); the Massachusetts Institute of Technology (MIT) Institute for Soldier Nanotechnology for nanoscale structure fabrication and material synthesis; Sandia National Laboratory for atomistic and mesoscale simulation methods development, and for precision material synthesis; the University of Illinois for nanoscale material diagnostics and atomistic material modeling; Northwestern University for material synthesis, material diagnostics, and atomistic simulations; and Texas A&M University for heat transfer characterization of periodic micrometer-scale structures. ERDC's interactions with the nanotechnology industry continue to grow as they partner in the development of technologies. A key area of interaction is the identification of impediments for rapidly advancing nanotechnologies through the acquisition process for Army applications with life cycle in mind. Industry partners with ERDC include Intelligent Automation, Inc., of Rockville, MD, and Nanosafe, Inc., of Blacksburg, VA.

DOE, NSF: EERE SETP continues to fund the NSF–DOE co-sponsored Quantum Energy and Sustainable Solar Technologies (QESST) Engineering Research Center (ERC), which conducts photovoltaic research that is relevant to nanotechnology (also applies to Objectives 1.3 and 2.2).

IC/DNI, NASA, DOD, universities, industry: In 2013, IC/DNI, NASA, and the DOD Air Force Space Command's Space and Missile Center will co-sponsor the 3rd Carbon Nanotubes for Space Applications Technical Interchange Workshop, along with Aerospace Corporation. The purpose of the workshop— previously attended by as many as 160 scientists, engineers, and program managers from Federal Government, industry, and academia—is to identify solutions to technical barriers for the successful manufacturing and use of bulk carbon nanotube materials in structures, sensors, electronics, energy generation, and memory storage systems. The third workshop will discuss recent research results on addressing these technical barriers (also applies to Objective 2.2).

NIH, USDA/NIFA: The workshop entitled —Using Nanotechnology to Improve Nutrition through Enhanced Bioavailability and Efficacy" was jointly sponsored by multiple NIH institutes and USDA/NIFA in 2012. Scientists from academic programs, food industries, Federal agencies, and non-governmental organizations were in attendance. The workshop assessed the state of knowledge and identified research gaps about the application of nanotechnology for delivering essential nutrients and other bioactive food components to promote health, and to encourage research using this nanotechnology to identify molecular targets for food components. Collectively, various opportunities were identified to use nanoparticles to influence absorption, distribution, metabolism, and/or elimination to promote health and disease prevention through the delivery of safe, targeted, and controlled interventions. The group also explored ways to enhance collaborations and to stimulate ideas for using nanotechnologies as they relate to enhancing nutrition and health. NIH and USDA program staffs are actively exploring strategies for promoting the application of nanotechnology for nutrition research and its translation (also applies to Objective 1.3).

NIST, DOE, industry: Working with SBIR-supported companies, NIST and DOE are jointly developing synchrotron spectroscopy imaging instruments and methods to characterize the chemistry, structure, and electronic states of engineered nanomaterials (ENMs). These instruments will be transferred to the new DOE synchrotron facility NSLS-II in 2014, providing enhanced capability of NIST's instruments for nanoscale spatial and depth resolution.

USDA/FS, industry, universities: FS continues to advance cellulose nanomaterials R&D projects in collaboration with multiple university partners in the Agenda 2020-led public–private partnership. The PPP advances cellulose nanomaterials R&D in forest products under a common vision developed with industry. University partners in the PPP are Pennsylvania State University, North Carolina State University, Georgia Tech, University of Tennessee at Knoxville, University of Maine at Orono, University of Minnesota, Oregon State University, Purdue University, and Mississippi State University.

Objective 1.2: Develop at least five broad interdisciplinary nanotechnology initiatives that are each supported by three or more NNI member agencies and support significant national priorities.

Coordinated Agency Contributions to Objective 1.2

Five NSIs have been developed to support significant national priorities, as listed below with collaborating agencies. A strategic summary of each NSI is provided in Chapter 4.

Sustainable Nanomanufacturing—Creating the Industries of the Future: DOD, DOE, EPA, IC/DNI, NASA, NIH, NIOSH, NIST, NSF, OSHA, and USDA/FS.

Nanotechnology for Solar Energy Collection and Conversion: DOD, DOE, IC/DNI, NASA, NIST, NSF, and USDA/NIFA.

Nanoelectronics for 2020 and Beyond: DOD, DOE, IC/DNI, NASA, NIST, and NSF.

Nanotechnology Knowledge Infrastructure (NKI): Enabling National Leadership in Sustainable Design: CPSC, DOD, EPA, FDA, NASA, NIH, NIOSH, NIST, NSF, and OSHA.

Nanotechnology for Sensors and Sensors for Nanotechnology: Improving and Protecting Health, Safety, and the Environment: CPSC, DOD/DTRA, EPA, FDA, NASA, NIH, NIOSH, NIST, NSF, and USDA/NIFA.

Objective 1.3: Identify and support goal-oriented nanoscale science and technology research aimed at national priorities informed by active engagement with academia, industry, and other stakeholders.

Individual Agency Contributions to Objective 1.3

DOD: The Navy is exploring extraordinarily lightweight materials with outstanding mechanical properties. Researchers at various laboratories are developing techniques for fabricating multiscale, hierarchical structures that not only are ultralightweight (<0.1 g/ml), strong, stiff, and damage-tolerant, but also affordable and scalable. The method involves using a photolithography process to create a polymer network of trusses that are then electroless-plated with a nanoscale metal. The final step is to remove the polymer, leaving a network of hollow tubes with nanoscale wall thickness. The hierarchical structure spans the range from —ano/micro" (wall thickness), micro/millimeter (truss length), to centimeter (material dimensions). This work has been featured in *Science* magazine (November 2011).

Exploration of application impacts of emerging nanomaterials work is of principal importance in the AFRL research portfolio. Nanomaterials and nanodevice research is informed by advances in nanomanufacturing to determine where application goals can be met. Nano-optical materials research is focused on quantitative improvements to filter technology and optical detector technology by exploring quantitative improvements in performance through novel zero-index nanomaterials and nanodevice enhancements of infrared detectors, including increased operating temperature through junction size reduction facilitated by transformation optics in a nanopatterned system. Evaluations of technical progress have been made in comparison with conventional materials approaches to these technical challenges. The teams executing these efforts include academic–industrial partnerships. The specific information has been used both for comparison with technical goals in these areas and to evaluate the overall engineering capability currently displayed by nanomaterials (also relevant to Objective 4.1 through the required evaluation of new process hazards).

Current approaches to bioengineering rely on an ad hoc, laborious, trial-and-error process, which can lead to long and costly development cycles that often do not translate to new designs. Transforming biology into an engineering practice would enable on-demand production of new and high-value materials, devices, and capabilities that have the potential to free warfighters from supply chains and to provide custom, distributed, on-demand manufacturing of materials in the field or on base. DARPA-sponsored research began in 2012 on new advanced biological tools and technology focused on design aids and databases; standardized, well-characterized modular parts and devices; test platforms and standardized chassis; low-cost and rapid DNA synthesis and assembly techniques that produce up to mega base-pair lengths of DNA; and tools for tracking and debugging the operation of synthetic gene networks. A key goal of the research is to engage industry and technology startups early in the development cycle and to combine with academic research to provide launching points for effective public–private partnerships (also relevant to Objective 2.2).

The Chemical and Biological Defense program has initiated new goal-oriented innovation programs in the areas of (1) dynamic responsive protective materials and (2) nanostructured therapeutic delivery vehicles. These new programs constitute competitive multiyear investments of several million dollars per year in product-focused multidisciplinary translational teams that span industry, academia, and government labs. For example, the composition of one of these teams, working on the project "Dynamically Responsive Chemical Biological (CB) Agent Defense System," spans two small businesses, three universities, two medical research institutes, and one DOD laboratory, with expertise that includes medical biology, nanoassembly, and textile manufacturing (also applies to Objectives 1.1 and 2.2).

DOE: The SunShot program within DOE/EERE has supported a variety of projects that use nanotechnology to drive down the cost of solar power installations. Approaches to reach the SunShot target of $1/W using nanotechnology include quantum dot solar cells, nanostructures for bandgap engineering, monolayers to adjust a material's work function or to passivate surfaces, and use of nanostructured materials for plasmonics and anti-reflective coatings. The majority of this research effort is supported by the Transformational PV Science and Technology: Next Generation Photovoltaics II (NextGen PVII) and the 2012 SunShot Concentrating Solar Power (CSP) funding announcements.

The solid-state lighting (SSL) program in DOE/EERE supports research and development of promising SSL technologies through annual competitive solicitations. Core Technology Research projects focus on applied research for technology development, with particular emphasis on meeting efficiency, performance, and cost targets. Product development projects focus on using the knowledge gained from basic or applied research to develop or improve commercially viable materials, devices, or systems. The research currently centers on light-emitting diodes (LEDs) and organic light-emitting diodes (OLEDs). Research topics include such items as emitter material research at the nanoscale level to reduce current droop and thermal sensitivity in LEDs. Also included is materials research and light extraction approaches for OLEDs. Manufacturing R&D projects accelerate SSL technology adoption through manufacturing improvements that reduce costs and enhance quality of SSL (also applies to Objective 2.1).

The Hydrogen and Fuel Cell Technologies Program (HFCT) in DOE/EERE conducts nanomaterials research and development for fuel cell, hydrogen production, and hydrogen storage applications. These efforts, such as those in catalyst R&D, will lead to lower costs and improved performance in hydrogen and fuel cell technologies and will help to enable widespread commercialization of these technologies.

NIH: NIH supports multiple goal-oriented nanoscale science and technology research aimed at national priorities that are informed by engagement and interaction across academia, industry, and other stakeholders. These are detailed in other sections of this report:

- For activities related to the SBIR/STTR programs, please see the end of Chapter 2.
- For activities related to the Translation of Nanotechnology in Cancer (TONIC) consortium, please see Goal 2, Objective 2.2.

NIST: As part of the national Materials Genome Initiative, NIST is developing models, validated with experimental data, to predict nanoscale mechanical properties, electrical properties, and structure of nanoparticle electronic materials.

NIST is making significant advances in nanoelectronics by advancing methods to separate, purify, and characterize CNTs and by elucidating key properties of graphene and graphene-based devices; developing nanowire-based test structures to advance high-performance transistor and nanophotonic applications;

improving processes to assemble molecular electronic devices with metal or silicon electrodes; finding new methods to understand spin transport, dynamics, and related magnetic interactions in electronic and data storage devices; harnessing our understanding of spin to characterize charge-based nanoelectronic structures; developing novel ways to measure and manipulate photons in nanophotonic applications; and coupling quantum optical and nanomechanical systems in silicon to create innovative devices for metrology and quantum communications applications. In addition, NIST is making significant advances in characterizing the energy barriers at materials interfaces that are critical for future ultralow-powered electronics.

NIST is advancing national priorities for advanced manufacturing by using nanotechnology to improve the safety of industrial robots and to create nanometrology sensors and devices for next-generation robotics and automation.

USDA/ARS: The Agricultural Research Service (ARS) investigates the use of nanomaterials to improve the properties and performance of value-added biobased products. One example is the use of methylcellulose and chitosan nanoparticles to produce healthy, edible films from fruits and vegetables. Another example is blow- and electro-spinning biopolymers to produce nanocomposites, nanofoams, and nanofibers. Other examples include the use of collagen-based nanofibers to produce biodegradable, high-efficiency air filters, and the development of pectin-based nanoparticles for controlled release of bioactive food ingredients in active packaging. In addition, ARS has developed nanocoatings to make flame-resistant cotton yarn, and it is depositing enzymes and peptides on cotton nanocrystals for potential applications in biosensors, antimicrobials, and bioremediation.

ARS is also investigating the application of nanotechnology to enhance food safety. Examples include the use of nanoscale biopolymer substrates with surface-enhanced Raman scattering (SERS) and of nanoscale DNA and RNA aptamers to detect food-borne pathogens and toxins. ARS is also researching zinc oxide and magnesium oxide nanoparticles to determine the mechanism of action behind their antimicrobial activity against foodborne pathogens.

USDA/FS: Successful FS-Agenda 2020 PPP research on cellulose nanomaterial research will create new high-value applications of wood fiber. This will in turn create manufacturing jobs in rural communities; it will also create new value for landowners. Creation of new jobs and creation of new value for landowners will incentivize forest managers to maintain the health of forests and invest in forest conservation and restoration.

Coordinated Activities with Other Agencies and Institutions Contributing to Objective 1.3

DOE, NSF: EERE/SETP has coordinated with NSF to fund the Foundational Program to Advance Cell Efficiency (FPACE) that continued through 2012. Combining both the technical and funding resources of DOE and NSF, this research investment aimed to reduce the gap between the efficiencies of prototype cells achieved in the laboratory and the efficiencies of cells produced on manufacturing lines. The FPACE program supports projects that are related to nanotechnology such as efforts to understand grain boundaries in solar cells using advanced measurement capabilities and computational analysis. EERE/SETP continues to fund the NSF and DOE co-sponsored QESST ERC, which conducts photovoltaic research.

DOE's Hydrogen and Fuel Cells Technology (HFCT) program activities are coordinated primarily with DOE/SC and NSF researchers, who work to understand the fundamental science behind these nanomaterials. The findings from the fundamental work are then applied to the HFCT R&D efforts to further develop and then to commercialize nanoscale technologies in hydrogen and fuel cells.

NIST, industry: As part of the joint NIST-Intel Emerging Nanoscale Interface and Architecture Characterization Program, NIST is developing methods to determine the nanoscale mechanical properties and stress states of nanoelectronic structures.

NIST, NNI agencies, international standards organizations: Through engagement in international standards organizations, NIST has actively coordinated with other national standards bodies, research organizations, and industries to develop validated methods to measure the properties of nanomaterials with consistent, verifiable results (also applies to Objective 2.5).

Objective 1.4: Develop quantitative measures to assess the performance of the U.S. nanotechnology R&D program relative to that of other major economies, in coordination with broader efforts to develop metrics for innovations.

Individual Agency Contributions to Objective 1.4

NIH: NIH institutes continually evaluate their strategic plans to assess the efficacy and diversity of funding across the agency. Specific to nanotechnology, NCI has initiated a program to develop new measures of performance that correlate the degree of collaboration and network integration to the levels of innovation, commercialization, and translation. This is a small effort ($50,000) to develop a metric of center-based research efficacy and efficiency. This initiative has the potential to grow into a much larger project tied to programmatic evaluation and the impact of nanotechnology on the biomedical field, with transNIH comparison, and a renewal effort in late 2013 and 2014 (also relevant to Objectives 1.1–1.4, 2.2–2.3, and 3.1–3.3).

NIST: As part of its plan for accelerating technology transfer and commercialization of Federal research in support of high-growth businesses, NIST will improve the collection of metrics relevant to nanotechnology, including participation in documentary standards committees, the impact of standard reference materials, the number of patents and licenses, the number and quality of publications, and the number of researchers participating in projects at the Center for Nanoscale Science and Technology (CNST), NIST's nanotechnology user facility.

USPTO: The U.S. Patent and Trademark Office (USPTO) of the DOC contributes a variety of nanotechnology-related patent data, which is used as a benchmark to analyze nanotechnology development and for trend analysis of nanotechnology patenting activity in the United States and globally. In March 2012, USPTO supplied information to the National Research Council for the Triennial Review of the NNI. The data in that presentation showed that since the start of the NNI, the percent of nanotechnology-related U.S. patents assigned or owned by universities has increased slightly, and the percent of nanotechnology-related U.S. patents with government interest statements has almost doubled. Both of these increases are likely attributable to the increased Federal R&D funding from agencies associated with the NNI. Regarding international nanotechnology-related patenting activity, U.S. inventors continue to lead in the total number of nanotechnology-related patent publications globally. However, that lead has been shrinking. U.S. inventors more significantly lead in the total number of nanotechnology-related patent publications in three or more countries, indicating a more aggressive pursuit of international intellectual property protection (also applies to Objectives 2.4 and 2.5).

Coordinated Activities with Other Agencies and Institutions Contributing to Objective 1.4

DOS, NNI agencies, OECD: The NNI and the Organisation for Economic Co-operation and Development (OECD) Working Party on Nanotechnology (WPN) co-organized the International Symposium and Workshop on Assessing the Economic Impact of Nanotechnology, held 27–28 March 2012 in Washington,

DC. The purpose of this meeting was to explore the need for and development of methodologies to assess the economic impact of nanotechnology across whole economies, factoring in many sectors and types of impact, including new and replacement products and materials, markets for raw materials, intermediate and final goods, and employment and other economic impacts. Principal financial support for this activity was provided by DOS. Presentations, video of plenary sessions, and four background papers are available on the NNI website at http://nano.gov/node/729.

NIST, other agencies: The Interagency Working Group on Technology Transfer, chaired by NIST, is working to identify opportunities for improving technology transfer from Federal laboratories, including nanotechnologies. This group also serves as a mechanism for information exchange and coordination with other agencies (also applies to Objective 2.2).

Goal 2: Foster the transfer of new technologies into products for commercial and public benefit

This NNI goal seeks to broaden and deepen the U.S. emphasis on commercial viability of nanotechnology through development of both practical applications and cost-effective scale-up of nanotechnology-based products and processes. These efforts will capture real benefits for national security, economic development, and job creation. The goal encompasses five objectives[11] that detail how the NNI will increase its investment, focus its resources, and broaden its engagement with academia, industry, and the international community to reach this goal.

The NNI member agencies have a number of activities uniquely targeting technology transfer and commercialization, for example, workshops to obtain input from industry and academia, SBIR and STTR programs to fund innovations in small businesses, and cutting-edge research infrastructure for use by all nanotechnology researchers, including those from industry.

Objective 2.1: Develop robust, scalable nanomanufacturing methods necessary to facilitate commercialization by doubling the share of the NNI investment in nanomanufacturing research over the next five years.

Individual Agency Contributions to Objective 2.1

DOD: The Navy is investigating a process for depositing very thick layers of nanostructured metals with unprecedented adherence, hardness, and corrosion resistance for application to ship structures, with potential for great reduction in maintenance and total ownership cost. The process, based on high-voltage, pulsed electrodeposition, is being applied under the Maintenance Free Ship program to produce main propulsion shafting that will last the life of the ship itself. Currently, wear and corrosion of shafts is one of the most important factors limiting the acceptable ship availability interval between major repair and overhaul.

The Defense Production Act Title III Program is establishing the infrastructure for the world's first manufacturing production facility for CNT yarn and sheet material. Emphasis is being placed on expanding flexible, scalable, and modular production processes; improving product quality and yield; and reducing manufacturing costs. Carbon nanotubes exhibit extraordinary strength and unique electrical properties and are highly efficient thermal conductors. They are the strongest and stiffest materials known in terms of

---

[11] For the full descriptions of the objectives of this NNI goal, see the 2011 *National Nanotechnology Initiative Strategic Plan,* http://nano.gov/node/581 or http://nano.gov/about-nni/what/vision-goals.

tensile strength and elastic modulus, respectively. CNT materials conduct electricity, shield from electro-magnetic interference and electromagnetic pulses, and enhance ballistics protection, while being impervious to corrosion, heat, or sunlight degradation. CNT yarn and sheet material can operate in a much broader temperature envelope than conventional materials. More importantly, the capability of Nanocomp Technologies, Inc., to produce these materials in both yarn and sheet formats enables turnkey insertion into customer system processes, reducing waste and time. Nanocomp is able to produce sheets as large as 4.5 ft. by 8.5 ft., the largest and only commercial CNT sheets reported in the world (also relevant to Objective 2.2).

DOE: In order to transfer new technologies into the market, EERE/SETP supports the SunShot Incubator Program. The Incubator Program provides early-stage assistance to help startup companies cross technological barriers to commercialization while encouraging private sector investment. Projects that relate to nanotechnology include concepts such as silicon nanowires to increase power density and reduce costs. Additionally, SETP funds three manufacturing development centers through the Photovoltaic Manufacturing Initiative (PVMI), along with the Process Development Integration Laboratory (PDIL) of the National Renewable Energy Laboratory (NREL).

The EERE solid-state lighting program supports research and development of promising SSL technologies through annual competitive solicitations. Manufacturing research and development projects accelerate SSL technology adoption through manufacturing improvements that reduce costs and enhance quality of LEDs and OLEDs. Research includes the development of high-speed, high-resolution, nondestructive test equipment with standardized test procedures and appropriate metrics within each stage of the value chain for semiconductor wafers, epitaxial layers, LED dies, packaged LEDs, modules, luminaires, and optical components. Also included is the development of manufacturing equipment enabling high-speed, low-cost, and uniform deposition of state-of-the-art OLED structures and layers.

IC/DNI: The National Reconnaissance Office has made significant progress toward Objective 2.1 (progress also relevant to Objective 2.2), including the following examples:

- CNT Memory—Development of nonvolatile low-power memory based on carbon nanotubes has entered the final phase of commercialization with several domestic manufacturers.
- CNT cables—In process of qualifying radio frequency coax cables, data/telemetry cables for space usage, and power cables that have achieved 1/15th the conductivity of copper power cables.
- CNT sheets—These are used to create ultralightweight composites, electromagnetic interference shielding, electrostatic discharge protection of commercial and military aircraft/spacecraft/launch vehicles, and lightning protection of aircraft.
- CNT yarns—These are used to create multifunctional hybrid composites, certain wire and cable forms, and communication accessories.
- CNT tape—This will be used for certain wire and cable forms and automated composite tape placement systems. Plans to demonstrate this with DOD during calendar year 2013 and 2014.
- Solar cells—Advanced III-V semiconductor materials have been used to achieved 35.1% efficiency in solar cells. Plans are now to pursue 41% efficient solar cells and scale for production the 35.1% cells.
- CNT additives for batteries—CNT additive research has been transferred from university research to a commercial domestic lithium ion battery manufacturer.

NASA: NASA continued to grow its investments in nanomanufacturing in 2012 and expects to continue this trend into 2013. A new effort in CNT-based high-strength reinforcements for composites was initiated in 2012 under the Space Technology Program. NASA is also collaborating with DOD and other NNI member agencies under the National Additive Manufacturing Innovation Institute formed in August 2012, and it is working on additive manufacturing of aerospace structures from a variety of materials, including CNT-reinforced high-temperature polymers.

NIH: NIH is investing in the translation of nanoscience discoveries into technologies that can facilitate the synthesis, characterization, and quality assurance in manufacturing or use of nanotherapeutics. This includes techniques, assays, and tools and devices for potential therapeutics.

NIOSH: NIOSH launched an initiative in 2012 to develop and disseminate case studies that demonstrate the utility of applying Prevention through Design (PtD) principles to nanomanufacturing. One element of the PtD program is the development of environmental, health, and safety (EHS) practices for incorporation into the texts and curricula of schools of engineering in the United States. In August 2012, NIOSH sponsored a workshop on Safe Nano Design to seek input into incorporating safe design into the synthesis, scale-up, pilot production, and full commercialization of nanotechnology-based products. The workshop was held in collaboration with the College of Nanoscale Science and Engineering, Albany, New York. Guidance and strategies developed at the workshop will be disseminated throughout 2013.

NIST: Compared with 2012, the request for 2014 will increase the intramural research budget for nanomanufacturing by nearly 60%. Measurement methods, standards, and technology are being developed to advance the understanding of fundamental processes, to enable efficient and sustainable production, and for high-throughput, closed-loop process control in the nanomanufacturing environment, with emphasis on carbon nanomaterials and roll-to-roll manufacturing processes. NIST nanomanufacturing-related programs are also addressing issues affecting materials lifetime, reliability, and life-cycle analyses.

NRC: NRC is monitoring developments in nanotechnology that might be applied within the nuclear industry in order to carry out its mission to ensure the safe use of radioactive materials for beneficial civilian purposes while protecting people and the environment (see http://www.nrc.gov/about-nrc.html). NRC staff participated in the June 2012 NanoNuclear Workshop organized by the Department of Energy and The Materials Society (see http://www.tms.org/meetings/2012/nanonuclear/home.aspx).

NSF: In 2013, NSF is continuing its contributions to translational innovation programs, including Grant Opportunities for Academic Liaison with Industry (GOALI), Industry/University Cooperative Research Centers (I/UCRC), the Innovation-Corps (I-Corps) program, Accelerating Innovation Research (AIR), Partnerships for Innovation (PFI), and PFI's Building Innovation Capacity (BIC) component. The NSF SBIR program has an ongoing nanotechnology topic with subtopics for nanomaterials, nanomanufacturing, nanoelectronics and active nanostructures, nanotechnology for biological and medical applications, and instrumentation for nanotechnology.

NSF will continue to support five NSECs (Nanoscale Science and Engineering Centers) that focus on manufacturing at the nanoscale. Those centers and the National Nanotechnology Infrastructure Network (NNIN) have strong partnerships with industry, national laboratories, and international centers of excellence, which put in place the necessary elements to bring discoveries in the laboratory to real-world, marketable innovations and technologies. The NSECs with a focus on nanomanufacturing are the University of Massachusetts at Amherst Center for Hierarchical Manufacturing (CHM); the University of California–Berkeley Center for Scalable and Integrated Nanomanufacturing (SINAM); the Northeastern University Center for High-Rate Nanomanufacturing (CHN); the University of Illinois at Urbana-Champaign Center for Nano-Chemical-Electrical-Mechanical Manufacturing Systems (Nano-CEMMS); and the Ohio State University Center for Affordable Nanoengineering (CAN).

Three Nanosystems Engineering Research Centers (NERCs), with a total five-year budget of approximately $55 million, were established in September 2012 and are starting to fully operate in 2013. Partnerships of new NERCs with small businesses in the areas of nanomanufacturing and commercialization will be strengthened while maintaining about the same level of NSF investment (also applies to Objective 2.2).

USDA/FS: In 2012, the FS Forest Products Laboratory formally commissioned a 30 kg/batch cellulose nanocrystal (CNC) pilot plant. A second 1 ton/day pilot plant to produce cellulose nanofibril (CNF) will be formally commissioned at the University of Maine in 2013. The pilot plants will produce sufficient quantities of CNC and CNF for government, academic, and industrial research. The pilot plants will also provide scientists with an opportunity to initiate manufacturing-related research at a scale larger than that of conventional laboratory batches.

Coordinated Activities with Other Agencies and Institutions Contributing to Objective 2.1

IC/DNI, DOD, NIST, NASA, industry, universities: Agencies involved in the Sustainable Nanomanufacturing Signature Initiative are coordinating with industry to create a consortium to accelerate the development of bulk carbon-nanotube-based multifunctional structures and manufacturing methods; demonstrate bulk CNT material systems and processes for nanomanufacturing roll-to-roll bulk CNT yarns, sheets, and tapes for multifunctional structures; and transition these materials and structures to CNT-based products manufactured in the United States (also relevant to Objectives 1.2, 1.3, and 2.2). Please refer to Chapter 4 for more details on this signature initiative.

NIH, FDA, NIST: The mission of the NCI-FDA-NIST Nanotechnology Characterization Laboratory (NCL, described in greater detail in Objective 2.2) includes improving synthesis of cancer nanomedical formulations to enable scale-up manufacturing as a part of commercial development.

NIST, industry: NIST is performing cooperative research and development with Applied Nanostructured Solutions (a Lockheed Martin Company), using advanced, *in situ* microscopy methods to evaluate the effects of varying synthesis conditions on the morphology and structure of CNT composites.

NSF, DOD, NIST, industry: NSF supports the National Nanomanufacturing Network (NNN), which includes the NSF NSECs and non-NSF centers in collaboration with DOD, NIST, and industry partners in an alliance to advance nanomanufacturing strength in the United States.

Objective 2.2: Increase focus on nanotechnology-based commercialization and related support for public–private partnerships.

Individual Agency Contributions to Objective 2.2

DOD: The Air Force Research Laboratory Manufacturing Technology program is investigating a lead-free solder system, addressed by nanotechnology development, to replace conventional lead/tin solder alloys. A key requirement is compatibility with current production environments, processes, and assembled components. In an AFRL Phase II program to develop lead-free solder based on nanocopper, through a process invented by Lockheed Martin, scientists have focused on primary process integration and commercialization issues. Process compatibility, materials production, and manufacturing efficiency have been explored in the course of this program. This activity has culminated in a demonstration of a representative electronic subassembly board. The work was performed at Lockheed facilities in order to improve transition potential for the technology (also relevant to Objective 4.4).

NIH: NIH is active in supporting partnerships to achieve commercialization and translation of nanotechnologies for biomedical applications. The NCI TONIC consortium, initiated by the NCI Alliance for Nanotechnology in Cancer, is an ongoing public–private partnership motivated to elevate awareness of academic and small-business biomedical cancer nanotechnology developments within the pharmaceutical industry and to enable precompetitive project support of science not fundable by NCI. In 2012, TONIC grew to over 25 members, including six large pharmaceutical companies and three advocacy groups. The TONIC Consortium has held quarterly meetings and has hosted a first-of-its-kind workshop on the Enhanced Permeability and Retention effect to address its utility in targeting nanomedicine to human cancer. The consortium is expected to continue to grow and drive the field into 2013. The TONIC activities contribute significantly to the following objectives within Goal 2:

- Objective 2.1: Sharing best practices in manufacturing nanomedical products.
- Objective 2.2: Enabling technology transfer from academic groups to small business and large pharmaceutical companies.
- Objective 2.4: Increasing exposure of small nanomedical companies to each other and to large pharmaceutical companies.
- Objective 2.5: TONIC membership includes international companies.

The NCI-sponsored Nanotechnology Characterization Laboratory has developed a comprehensive assay portfolio for assessing the safety of nanoparticles used in *in vivo* applications and has characterized to date over 250 different nanoparticle formulations while contributing data to several IND applications with FDA. The National Institute of Environmental Health Sciences (NIEHS) and the National Toxicology Program (NTP) are focused on the assessment of properties relevant to acute and chronic exposures of workers.

The NCI Alliance for Nanotechnology in Cancer is focusing on developing and translating nanotechnology-based techniques, assays, and tools and devices for early disease diagnosis, and for *in vivo* imaging techniques, in addition to the delivery of new therapies. The Alliance academic investigators have been prolific in leveraging government funds and raising additional funds from philanthropic and private sector sources to achieve these goals. The alliance currently has over 50 companies (mostly spinoffs) associated with its operation and 10 ongoing clinical trials in the areas of diagnostics and therapy. Examples of these efforts and clinical trials include:

- Systemic delivery of chemotherapeutics using polymeric nanoparticles.
- Nanoparticle-based delivery of small-interfering ribonucleic acid (siRNA).
- Protein detection and microfluidics technologies for early diagnosis and therapy monitoring.

USDA/FS: The FS FPL will continue to partner with an Agenda 2020-led PPP in cellulose nanomaterials research. Collaborative projects in the PPP have been developed under a common vision to research precompetitive science and technology for the purpose of commercializing cellulose nanomaterial-based products. There are nine university partners in the PPP. At present, cellulose nanomaterials research projects under the PPP include predictive modeling of cellulose nanomaterials, composites, sensors, aircraft materials, and cellulose nanomaterials production. The PPP was formally funded in 2011.

Coordinated Activities with Other Agencies and Institutions Contributing to Objective 2.2

DOE, NSF: EERE/SETP supports the NSF and DOE co-sponsored QESST Engineering Research Center to work with companies to advance technologies toward commercialization.

IC/DNI, DOD, industry: Nanomanufacturing of bulk carbon nanotube sheets and yarns has attracted large commercial investment from DuPont ($10 million), and the Office of the Secretary of Defense (OSD) Defense Production Act Title III has added an additional $3.2 million appropriated in 2012. These funds will assist in building out the world's first bulk carbon nanotube production pilot plant in Merrimack, NH.

NIST, NSF, industry, universities: In 2013, NIST and NSF will continue support for the Nanoelectronics Research Initiative (NRI). The NRI is a public–private partnership supporting the development of future computing devices, including collaborative research between NIST and NRI researchers from industry and from NSF-funded universities. NIST's 2013 contribution to the NRI is via a new award for up to five years. NSF has separate agreements with the NRI and the Semiconductor Research Corporation (SRC) to facilitate research and education cooperation between the NRI and over 20 university activities funded by NSF, including centers and networks (NSECs, NERCs, Materials Research Science and Engineering Centers, and the Network for Computational Nanotechnology/nanoHUB). NSF also co-sponsors workshops on future trends in nanoelectronics in partnership with NRI and with the SRC. NIST has also supported SEMATECH and U.S. universities by providing wafer-scale calibration methods for through-silicon vias to reduce interconnect delays and create three-dimensional integrated circuits.

NNI Agencies, other government members, industry, universities: The NNI Workshop on Regional, State, and Local (RSL) Initiatives in Nanotechnology, was held 1–2 May 2012 in Portland, Oregon. This was the fourth in a series of RSL workshops sponsored by the NNI since 2003.[12] RSL organizations have vital roles in establishing the infrastructure, preparing the skilled workforce, and supporting the industries—especially startups and small and medium enterprises—that are critical to building an economically viable nanotechnology innovation ecosystem in the United States (also relevant to Objective 2.4). Supporting partnerships with such organizations has been a goal of the NNI since its inception.

Objective 2.3: Establish and/or sustain national user facilities, cooperative research centers, and regional initiatives with the goal of accelerating the transfer of nanoscale science from discovery to commercial products.

Individual Agency Contributions to Objective 2.3

DOE: The Office of Science continues to operate five Nanoscale Science Research Centers (NSRCs), national user facilities for interdisciplinary research at the nanoscale, serving as the basis for a national program that encompasses new science, new tools, and new computing capabilities. Each center has particular expertise and capabilities in several selected theme areas, such as synthesis and characterization of nanomaterials; catalysis; theory, modeling, and simulation; electronic materials; nanoscale photonics; soft and biological materials; imaging and spectroscopy; and nanoscale integration.

EERE/SETP supports the National Renewable Energy Laboratory though an annual operating plan that contains a collection of efforts related to nanotechnology. For example, research utilizing nanomaterials is being conducted to enhance transparent conducting materials and to increase the resolution of measurement techniques to electrically and optically probe grain boundaries, interfaces, and structures. The BRIDGE program supports the development of advanced characterization and process development tools at a number of DOE/SC-funded user facilities to significantly lower the cost of solar energy. The program is the first to provide collaborative research teams from industry, universities, and national laboratories to work together on solar energy R&D in DOE's research facilities with the aim of more rapidly transitioning discoveries into products.

NIH: Translational outputs from the cooperative research Centers of Cancer Nanotechnology Excellence supported by NCI from 2012 include the following:

- Development and commercialization of integrated blood barcode chips to measure panels of biomarkers in glioblastoma patients being treated with Avastin to monitor therapeutic response and immunotherapy for melanoma.

---

[12] Previous workshops took place in 2003, 2005, and 2009. The 2003 and 2009 workshops produced formal reports, available at http://nano.gov.

- Development and clinical trials using micro-NMR (nuclear magnetic resonance) devices for rapid molecular analysis of human tumor samples and circulating tumor cells detection and profiling in whole blood.
- Development of materials and instruments for improved endoscopy allowing for specific identification of tumorous polyps through surface-enhanced Raman spectroscopy using tumor-targeted nanoparticles.
- Clinical trials involving gold nanoshells for localized thermal ablation in head and neck cancer and prostate cancer.

NIST: The NIST nanotechnology user facility, the CNST, has recently added and plans to add in 2014 new commercial tools in the nanofab specifically for the purpose of enabling researchers to fabricate prototypes at a production scale. A high-throughput i-line stepper is now available that can expose up to 120 wafers per hour, each with a diameter of 200 mm, with alignment accuracy of 40 nm, and overlay resolution down to 280 nm. Additional production-scale tools expected in 2014 include an automated sputter system, a deep silicon etcher, a photolithography track system, and additional electron-beam lithography capabilities.

Coordinated Activities with Other Agencies and Institutions Contributing to Objective 2.3

NIST, DOE, NSF, DOL, DOEd, NASA, DOD: The interagency Advanced Manufacturing National Program Coordination Office (AMNPO), hosted by NIST, was established in 2012 to create a whole-of-government advanced manufacturing initiative. In 2012 the AMNPO held four workshops to solicit public input on the design for a new program, the National Network for Manufacturing Innovation (NNMI), which was proposed in the President's 2013 Budget. The NNMI is envisioned as a network of institutes that will serve as regional hubs of manufacturing excellence, including in emerging areas such as nanomanufacturing.

Objective 2.4: Assist the nanotechnology-based business community, including small- and medium-sized enterprises, in understanding the Federal Government's R&D funding and regulatory environment.

Individual Agency Contributions to Objective 2.4

DOC/BIS: The Bureau of Industry and Security (BIS) advances U.S. national security, foreign policy, and economic objectives by ensuring an effective export control and treaty compliance system, and by promoting continued U.S. leadership in strategic technologies, including nanotechnology. BIS accomplishes its mission by maintaining and strengthening adaptable, efficient, and effective export control and treaty compliance systems. BIS export control outreach and education constitute the first line in the bureau's contact with U.S. exporters and provide guidance and transparency to new and experienced exporters regarding the Export Administration Regulations (EAR). BIS's activities include seminars, webinars, teleconferences, and on-location panel sessions at various conferences. Additionally, one-on-one counseling assistance is provided on both the East and West Coasts for extended hours of daily operation. Over the past few years, BIS has developed capabilities to offer training online and via interactive webinars. Through these programs, BIS provides guidance on regulations, policies, and practices and helps to increase compliance with U.S. export control regulations. BIS has also developed an introductory series of easy-to-use training modules. This service offers exporters and re-exporters a cost-saving mechanism to learn about U.S. dual-use export controls. BIS services are particularly useful for small- and medium-sized businesses that operate with limited compliance resources.

USDA/FS: FS assists the forest products industry technology alliance, Agenda 2020, in its research funding. Agenda 2020 develops and manages research projects that aim to create game-changing precompetitive technology, including cellulose nanomaterials for the pulp and paper industry.

USDA/NIFA: NIFA's diverse SBIR programs have continued to support nanotechnology R&D aiming at commercialization. The applications supported in 2012 have included nanotechnology-enabled sensor technologies for detection of microbial pathogens and insects and crop environmental stresses; innovative lighting for environmental chambers for growing vegetables; enhanced thermal processes; and using nanocellulose materials in polymer composites to improve functional properties of packaging (also applies to Objective 2.2).

USPTO: The transfer of new nanotechnology-related technologies into products for commercial and public benefit depends on effective mechanisms that protect new ideas and investments in innovation and creativity. USPTO is at the cutting edge of the Nation's provision of intellectual property policy advice and guidance to the Executive Branch. The agency has put in place several initiatives to keep pace with the rapid advances being made in nanotechnology. USPTO continues to host a nanotechnology customer partnership forum that provides a mechanism for USPTO officials and patent stakeholders to share concerns and information related to the patenting of nanotechnology. The agency continues to provide in-depth nanotechnology-specific training events for patent examiners as well as fostering communication among examiners across multiple disciplines to enhance the technical knowledge base of USPTO examiners. As a result of this ongoing training initiative, collaborative environment, and targeted hiring of patent examiners with specific nanotechnology experience, the USPTO has a subset of patent examiners across all technology disciplines who serve as points of contact to assist other examiners with nanotechnology issues related to patent examining.

Coordinated Activities with Other Agencies and Institutions Contributing to Objective 2.4

NNI agencies through the NILI Working Group: Led by NIST and DOD, members of the Nanomanufacturing, Industry Liaison, & Innovation (NILI) Working Group of the NSET Subcommittee, with strong support from NNCO, perform frequent outreach to the nanotechnology business community, including regional, state, and local alliances. This is accomplished through a number of outreach mechanisms, including presentations at regional and national meetings and regular stakeholder conference calls.

Objective 2.5: Increase international engagement to facilitate the responsible and sustainable commercialization, technology transfer, innovation, and trade related to nanotechnology-enabled products and processes.

Individual Agency Contributions to Objective 2.5

DOC/BIS: BIS has developed and published two comprehensive documents on "Export Management and Compliance Programs (EMCPs)," which are programs that can be established to manage export-related decisions and transactions to ensure compliance with the Export Administration Regulations and license conditions. Specifically, BIS has published "Compliance Guidelines: How to Develop an Effective Export Management and Compliance Program and Manual" and "EMCP Audit Module: Self-Assessment Tool." Additionally, BIS conducts regularly scheduled outreach seminars dedicated to the topic "How to Develop an Export Management and Compliance Program" and has added the webinar "Elements of an Effective Export Compliance Program" to its menu of online training offerings.

DOD: To seed innovation in areas that are relevant to Air Force capabilities and that can potentially benefit from research and technology development in nanomaterials and devices, the Air Force Research Laboratory has engaged the international community to encourage work in specific areas such as radio-frequency photonic nanodevices, nanomagnetic materials, and optical nanoantennas. The AFRL Materials and Manufacturing Directorate has established recognized metrics to facilitate wide technical evaluation on the part of researchers and both guide and identify research with technological potential in these areas. This will rapidly accelerate the innovation process, pairing emerging research and discoveries with technical venues, and speeding up the transfer of technology as a result. This process can also identify classes of materials that may provide unique advantages over a technology area, such as nanopatterned metallic structures, and allow for focused nanomanufacturing capability development to support broad-based commercialization.

USDA/FS: FS continues to lead international standards development for cellulose nanomaterials. In 2012, the FS-led cellulose nanomaterials terminology standards development committee of the Technical Association of the Pulp and Paper Industry completed three rounds of commenting on a proposed draft of standard terms and definitions for cellulose nanomaterials. The proposed draft will be balloted in the first or second quarter of 2013. FS will continue to lead the coordination of cellulose nanomaterials international standards development in the coming years.

USDA/NIFA: NIFA staff has actively engaged with international nanotechnology research communities and foreign government agricultural research agencies and academic institutes to foster collaborative and cooperative relationships between U.S. food and agricultural scientists and engineers and their foreign counterparts. NIFA staff has been instrumental in establishing the International Society of Food Applications of Nanoscale Science (ISFANS) with its mission —et strengthen research, communication, dissemination of information and networking for technology transfers and international collaborations among interested parties from academia, industry, government, consumers, and other participants around the world." In addition, the AFRI nanotechnology program has encouraged all applications, especially those with commercial impact in sight, to include economic analyses of anticipated benefits to agriculture, food, and society in its 2012–2013 combined RFA.

USPTO: The USPTO is moving to a new patent classification system, the Cooperative Patent Classification (CPC), which is a jointly shared and operated classification system used by USPTO and the European Patent Office (EPO). This new classification system is based on an internationally used classification system, the International Patent Classification (IPC), which contains most of the world's patent documents. As the USPTO moves to this new classification system, there will be a more harmonized, internationally consistent classification of nanotechnology-related patent documents.

Coordinated Activities with Other Agencies and Institutions Contributing to Objective 2.5

DOS, NNI Agencies, OECD: The OECD's Working Party on Nanotechnology provides advice on emerging policy issues of science, technology, and innovation related to the responsible development of nanotechnology. DOS serves as Vice-Chair of the WPN bureau and coordinates NNI agency participation in ongoing projects, including —Regulatory Frameworks for Nanotechnology in Food and Medical Products," —Green Growth and Nanotechnology," and —Nanomedicine Case Studies." For its 2013–2014 program of work and budget, the WPN will consider:

- Policies to understand and monitor the societal and economic impacts of nanotechnology.
- Linking nanotechnology to broader policies for sustainable and green growth.
- Policy issues raised by the convergence of nanotechnology with other technologies.

NIOSH, NNI agencies, European Union (EU): NIOSH, in collaboration with other NNI agencies, participates in multiple Communities of Research (CoRs) as part of the U.S.–EU effort to bridge EHS research. NIOSH contributes in the areas of human health, risk assessment, exposure assessment, and risk management and controls. In addition to active participation in the CoRs, NIOSH has actively engaged research institutes in Canada, Brazil, and Russia in several areas of nanotechnology EHS research collaboration.

NIST, NNI agencies, industry: In 2013, NIST co-chaired a meeting of the American National Standards Institute's Nanotechnology Standards Panel to determine if current nanotechnology standards activities meet existing stakeholder needs, and to assess the impact of existing nanotechnology standards on research and development. The outcome of this assessment will help lay the foundation required to ultimately achieve Objective 2.5.

Goal 3: Develop and sustain educational resources, a skilled workforce, and the supporting infrastructure and tools to advance nanotechnology

This NNI goal seeks to develop all aspects of the infrastructure necessary to the advancement of nanotechnology-based knowledge, products, and practices that will benefit the Nation and its citizens. The goal encompasses three objectives[13] that detail how the NNI will responsibly engage and educate the public and the workforce regarding the opportunities that nanotechnology offers and the skills it requires, along with providing the needed access to advanced facilities and tools.

Significant progress is being made on all three aspects of Goal 3 (developing and sustaining educational resources, a skilled workforce, and supporting infrastructure and tools). With respect to education and workforce development, education is among the chief objectives of NNI-funded university research. In addition, specific programs that target K–16 education, improve nanotechnology curricula in U.S. schools and universities, and educate the public about nanotechnology are growing in scale and reach. Development of the extensive network of research centers, user facilities, and other infrastructure for nanotechnology research, a key element of the original NNI strategy, is now largely complete, although there are ongoing investments to maintain the equipment and facilities at the cutting edge.[14]

Objective 3.1: Initiate, develop, support, and sustain programs for educating, training, and maintaining a skilled nanotechnology workforce.

Individual Agency Contributions to Objective 3.1

DOD: The Defense Threat Reduction Agency Basic and Applied Sciences program contributes to nanoscience education through its support of single- and multi-investigator research projects in academia and small business. For example, elements of a project related to interrogation of gold nanorod and polypeptide composites were incorporated into an undergraduate/graduate elective course entitled —Nanobiotechnology" developed by researchers at Arizona State University. Evidence of the quality of training provided, as well as the quality of participating students, is provided by recognition of the winning student poster —Electronic Stimuli-Responsive Hybrid Nanoparticle-Biopolymeric Materials" at the 2011 American Institute of Chemical Engineers National Meeting.

---

[13] For the full descriptions of the objectives of this NNI goal, see the 2011 National Nanotechnology Initiative Strategic Plan, http://nano.gov/node/581 or http://nano.gov/about-nni/what/vision-goals.

[14] For more information on NNI-supported research infrastructure see http://nano.gov/centers-networks.

DOE: By supporting fundamental and applied R&D, DOE supports students at both the undergraduate and graduate levels as well as postdoctoral researchers to gain expertise in nanotechnology. Additionally, the businesses supported by programs such as the SunShot Incubator and SBIR/STTR help create a skilled nanotechnology workforce. Furthermore, the Minority University Research Associates (MURA) program directly supports students working on concepts such as novel nanomaterials with optoelectronic properties tailored for PV cells and understanding the impact of nanoparticles on thermal energy storage.

The DOE Office of Science continues to operate five Nanoscale Science Research Centers, national user facilities for interdisciplinary research at the nanoscale. The NSRCs serve as the basis for a national program that encompasses new science, new tools, and new computing capabilities. The NSRCs provide training for graduate students and postdoctoral associates in interdisciplinary nanoscale science, engineering, and technology research.

FDA: In prior years, FDA personnel training has focused on course work, including lectures by experts in science and regulatory issues surrounding nanotechnology, and workshops, to provide laboratory training with analytical equipment used to characterize the physical and chemical properties of engineered nanomaterials relevant to FDA review of products that may contain nanomaterials or otherwise involve nanotechnology. In 2014, FDA will continue to provide such training, including laboratory training with analytical equipment used to characterize the physical and chemical properties of engineered nanomaterials. This type of training is invaluable to FDA staff members who review product applications that include data derived from such instrumentation.

IC/DNI: The National Reconnaissance Office has competitively selected 2012 Intelligence Community postdoctoral researchers to work on nanotechnology challenges to advance the state of the art of focal plane arrays, specifically dark current and nanopolymorphic materials ultrasensitive to ultraviolet light, for big data (up to 12 terabytes/disc) holographic storage media. The 2013 IC Post-Doctoral Broad Agency Announcement was released in December 2012. NRO has proposed two new nanotechnology research efforts: (1) nano-resin for ultrastrong, ultralightweight nanocomposites and (2) carbon nanotube chirality control and separation by chirality.

NASA: NASA recognizes that a pool of talented research scientists and engineers is critical to the continued economic well-being of the Nation and, in particular, the aerospace industry. In 2011, NASA initiated the Space Technology Research Fellowship Program to feed the pipeline of scientists and engineers to meet the future needs of the aerospace community. Eighty fellowships were awarded in 2011, with eight of these being in nanotechnology-related areas. Another 48 fellowships were awarded in 2012, seven in nanotechnology topics. In 2012 the Space Technology Research Grants Program initiated two new activities, Early Stage Innovation and Early Stage Career Awards, which funded grants to university faculty to perform fundamental research and develop technologies in support of future NASA mission needs. Five grants were made in nanotechnology-related topics, such as nanocomposites for radiation shielding, nanolattice-based multifunctional materials, nanocomposite membranes for water purification, and small satellite propulsion and steering concepts based upon optical metamaterials. NASA also supports a significant number of grants in nanotechnology through the Experimental Program to Stimulate Competitive Research (EPSCoR). In 2011, NASA EPSCoR funded about $3 million in grants in nanofluids for heat transfer, carbon-nanotube-reinforced 3D structural composites, nanomaterials for photovoltaics, and assessing the effects of space radiation on the performance of nanoelectronic devices. Six new awards were announced for 2012 in nanotechnology topics ranging from CNT-based high-strength fibers to 3D nanostructured solid-state capacitors. Similar levels of support for nanotechnology R&D at universities are expected in 2013 under the Space Technology Research Grant and EPSCoR programs.

NIH: Successful biomedical research depends on the talent and dedication of the scientific workforce. NIH supports many innovative training programs and funding mechanisms to foster scientific creativity and exploration. The goal is to strengthen our Nation's research capacity, broaden our research base, and inspire a passion for science in current and future generations of researchers. This process occurs through the career development fellowship training grants, academic institutional expansion, and specialized workshops. The NCI Alliance for Nanotechnology in Cancer will continue to support several training components through 2014. Six Cancer Nanotechnology Training Centers (CNTCs) are supported by the Alliance. These centers support graduate students, postdoctoral fellows, and medical graduates in dual-mentored environments structured to build transdisciplinary skills needed for future biomedical nanotechnologists. These CNTCs are building a collaborative network to share resources and to evaluate best practices.

NIOSH: Over the past five years, NIOSH has developed and delivered several levels of training for professional industrial hygiene and EHS practitioners. Included in this effort are technical symposia arranged by NIOSH; lectures and presentations by experts on the full range of occupational safety and health issues associated with nanotechnology; and conducting hands-on training on the basic techniques needed for exposure assessment and exposure control verification. Direct training of visiting scientists in NIOSH research laboratories has been offered in the areas of toxicology, aerosol characterization, and analytical measurements.

NIST: NIST is sustaining its efforts to create a skilled nanotechnology workforce through cooperative agreements with academic institutions supporting postdoctoral researchers and visiting fellows, centrally supported National Research Council postdoctoral research associates, and outreach activities such as lectures at the annual NIST Summer Institute for Middle School Science Teachers and an educational workshop on —Microscopy for Science, Technology, Engineering, and Mathematics (STEM) Educators" offered at the SPIE conference, —Scanning Microscopies: Advanced Microscopy Technologies for Defense, Homeland Security, Forensic, Life, Environmental, and Industrial Sciences."

NSF: Education-related activities include development of materials for schools, curriculum development for nanoscience and engineering, development of new teaching tools, undergraduate programs, technical training, and public outreach. NSF will continue to support the two networks, Nanoscale Informal Science Education Network (NISE Net) and Nanotechnology Applications and Career Knowledge (NACK) Network, for nanotechnology education with national outreach.

USDA/NIFA: Education is one of the three key pillars supporting NIFA's mission and programs. NIFA's higher education programs have been opened to proposals to develop new educational curricula in institutions of higher education and to improve capacity-building in minority-serving institutions and tribal colleges and universities. New sciences and novel ideas such as nanoscale science and nanotechnology have been supported in the NIFA educational programs.

Coordinated Activities with Other Agencies and Institutions Contributing to Objective 3.1

DOE, NSF: EERE/SETP and NSF co-sponsor the QESST ERC, which has a primary goal of creating educational resources and a skilled workforce in solar technology.

NIST, Army Corp of Engineers (ACE): NIST helped ACE develop competency in nanomechanical measurements by providing extended, on-site training to ACE personnel.

NIST, NSF: The jointly-funded Summer Undergraduate Research Fellowship (SURF) program provides nanotechnology research opportunities throughout NIST's laboratories and user facilities at both the

Gaithersburg, MD, and Boulder, CO, campuses; in 2012 over 30 SURF students participated in nanotechnology-related projects.

Objective 3.2: Initiate outreach and informal education programs and publish related information to foster a student population, workforce, and public that are well informed about the opportunities in nanotechnology-related industries and the potential impacts of environmental, health, and safety (EHS) and ethical, legal, and societal implications (ELSI) of nanotechnology.

Individual Agency Contributions to Objective 3.2

DOD: Through the Minority Leaders Fellowship Program, AFRL is reaching out to students nationwide and sponsoring summer internships and research at their labs and home institutions. The focus of this work in nanomaterials and devices allows for greater exposure of students to topics in both nanoscience and nanotechnology-related safety. This initial contact with AFRL leads to ongoing relationships between students and AFRL researchers. Return interns, fellowship candidates, co-op employees, and new hires have resulted from this program, which both bolsters the overall workforce familiarity with nanotechnology issues and renews the AFRL workforce. Projects in nano-optical coatings for improved performance in imaging systems, tunable nano-optical devices, and nano-modified fabrics for wearable electronics are among the projects that have showed promising results based on the efforts of student researchers. The experience gained by these students in both in the technical content of their work and in exposure to nanoscience laboratory practices strengthens their studies throughout their careers (also relevant to Objective 4.4).

The DTRA Basic and Applied Sciences program supports investigation of nanoscale phenomenology relevant to development of new materials, including those with prospective dual use for both the DOD and civilian medical communities. For example, the recent publication —Syergistic Administration of Photothermal Therapy and Chemotherapy to Cancer Cells using Polypeptide-based Degradable Plasmonic Matrices" (*Nanomedicine* 2011; 6:459-473) is derived from work on polypeptide-based nanomatrices, funded under the auspices of the Basic and Applied Sciences program for securing and monitoring WMDs.

NIH: During 2013, the National Institute of Nursing Research will convene an expert working panel on Nanotechnology for Advancing Nursing Science to highlight current knowledge of nanotechnology with respect to patient care, to elucidate potential applications of nanotechnology in addressing nursing research issues, and to discuss strategies for effectively incorporating nanotechnology into interventions for improved patient outcomes.

In addition to supporting cadres of graduate students and postdoctoral trainees, the nine Centers for Cancer Nanotechnology Excellence funded by the NCI Alliance for Nanotechnology in Cancer have a portion of their budgets dedicated to education, training, and outreach. These centers have supported a variety of courses, lectures, seminars, and community information forums to enhance knowledge of medical nanotechnology among both scientists and the public.

NIOSH: NIOSH maintains regular communication with academia and the public on its nanotechnology research program in the form of e-newsletters, the NIOSH Science Blog, technical meeting and symposia presentations, formal peer-reviewed journal publications, and other NIOSH publications. The NIOSH —Nanotechnology Topic" page is one of the most frequently visited on the agency website. A recent NIOSH initiative is focused on delivering knowledge and training to the industrial hygiene and EHS practitioner community. In 2012, NIOSH started delivery of a higher-level exposure measurements and assessment strategy development course to EHS professionals in the State of California, to the International

Occupational Hygiene Association (Kuala Lumpur), and to EHS professionals in Brazil in collaboration with Fundacentro.

NIST: NIST issues biweekly, widely distributed news articles written for the general public that frequently include nanotechnology-related topics, including EHS. In addition, CNST issues a quarterly newsletter through ₋GovDelivery" to over 4500 researchers, students, and members of the general public, providing highlights of nanotechnology research at CNST and information about new tools and capabilities available there.

USDA/NIFA: The AFRI nanotechnology program has supported informal education and extension outreach projects to contribute an enhanced perception and acceptance of nanotechnology applications related to food and agriculture. At its 2012 annual grantees meeting, a one-day symposium titled ₋Public Perception, Acceptance, Education and Outreach of Nanotechnology in Food and Agriculture" was held to assess the outcomes of the funded research projects and to promote discussion on the subject among experts and researchers. Broad societal implications and regulatory issues were presented by the invited experts. The participants also studied hands-on nanotechnology exhibits from Cornell University, funded by NSF and NIFA.

Coordinated Activities with Other Agencies and Institutions Contributing to Objective 3.2

NSF, EPA: NSF and EPA are engaged in an ongoing collaboration through a formal memorandum of understanding (MOU) to support two center-scale awards focused on nanotechnology-related environmental, health, and safety research and education, the Center for Environmental Implications of NanoTechnology at Duke University and the Center for Environmental Implications of Nanomaterials at UCLA. Over the past five fiscal years, EPA has provided a total of $5 million through interagency funding agreements to support the two centers, and NSF has provided $34 million. This joint funding from EPA and NSF supports the full breadth of EHS research and education activities at the centers, including informal education and outreach. It also supports research and education on the ethical, legal, and societal implications (ELSI) of nanotechnology.

NSF, NIH: The Museum of Science had a Nanoscale Informal Science Education (NISE) network award from NSF and an NIH Science Education Partnership Award (SEPA), which was partially spent to create a media-based interactive exhibit, the Nanomedicine Explorer. The NIH SEPA predated the NISE Net award, but the work it supported eventually benefited the NISE Net.

Objective 3.3: Provide, facilitate the sharing of, and sustain the physical R&D infrastructure for nanoscale fabrication, synthesis, characterization, modeling, design, computation, and hands-on training for use by industry, academia, nonprofit organizations, and state and Federal agencies.

Individual Agency Contributions to Objective 3.3

DOD: Army Research Laboratory (ARL) nanotechnology research focuses on the synthesis, characterization, and processing of nanoscale materials into novel composites for lightweight armor systems. One of the largest impediments to greater nanocomposite use is the propensity of nanomaterials to agglomerate or "clump" during processing, negating their unique properties. To surmount this obstacle, ARL is preparing to stand up a new laboratory focused on the safe and scalable production of nanocomposites, and to probe the phenomena leading to their efficient dispersion. ARL is currently exploring bio-derived nanocellulose materials as reinforcements for greener nanocomposites, and it is developing tools to allow for the measurement of nanoscale physical properties in soft materials.

DOE: The five Nanoscale Science Research Centers, which are operated by the Office of Science at DOE, are housed in custom-designed laboratory buildings near one or more other major DOE facilities for X-ray, neutron, or electron scattering, which complement and leverage the capabilities of the NSRCs. These laboratories contain cleanrooms, nanofabrication resources, one-of-a-kind signature instruments, and other instruments not generally available except at major user facilities. These facilities are routinely made available to the academic, industrial, and nonprofit research communities during normal working hours.

NIST: NIST is sustaining the significant increase made in 2012 in the annual expenditures devoted to recapitalizing and updating the CNST user facility's equipment and instrumentation (reported in PCA 6) in order to maintain the cutting-edge research environment required to meet the nanoscale measurement and fabrication needs of industry, academia, nonprofit organizations, and state and Federal agencies.

Coordinated Activities with Other Agencies and Institutions Contributing to Objective 3.3

NIST, NIOSH, CPSC, FDA, EPA, NIH (NCI, NIEHS): As part of the 2011 NNI Environmental, Health, and Safety Research Strategy, NIST is leading the development of a national measurement infrastructure for engineered nanomaterials. NCI's Nanotechnology Characterization Laboratory (NCL) is also collaborating with other agencies in the characterization of nanomaterials.

NSF, universities: NSF supports ongoing operations of the National Nanotechnology Infrastructure Network (NNIN), which connects 14 university nanotechnology research facilities. It is recompeting the network in 2013. NNIN serves about 6,100 individual users annually. Of these, about 5,100 are academic users, principally graduate students, undergraduates, and postdocs from about 200 academic institutions. Start-ups and small companies use NNIN facilities as prototyping labs, contributing to innovation in new technologies and initial steps to commercialization. In 2012 about 16% of users were from industry, involving over 360 small companies. Since the beginning of the NNIN in 2004, 79 small companies have been founded based on technology developed at NNIN facilities.

Goal 4: Support responsible development of nanotechnology

This NNI goal, of equal importance to the other three pillars of the NNI goal framework, integrates the pursuit of world-class nanotechnology development with sustainable and responsible research, development, and commercialization practices. The goal encompasses four objectives[15] that detail how the NNI will address safety and risk issues across the life cycles of nanomaterials; foster productive and ongoing domestic and international dialogues on best practices in nanotechnology development; proactively address ethical, legal, and social issues associated with nanotechnology development; and apply best practices at all levels to ensure that nanotechnology benefits society and the environment over the long term. These objectives are fully consistent with the research aims detailed in the 2011 NNI Environmental, Safety, and Health Research Strategy (http://nano.gov/node/681).

The NNI is making significant progress towards realizing the goal of supporting responsible development of nanotechnology. Funding for nanotechnology-related EHS research continues to increase at a rate far in excess of the overall NNI budget growth rate, growing from $35 million in 2005 to $121 million in the 2014 request. This funding only includes the narrowly defined *primary purpose* EHS R&D, although many other research activities also bear on EHS work. The NNI agencies have developed a comprehensive interagency strategy to move EHS investments forward effectively, in line with the roles and responsibilities of the

---

[15] For the full descriptions of the objectives of this NNI goal, see the 2011 *National Nanotechnology Initiative Strategic Plan*, http://nano.gov/node/581 or http://nano.gov/about-nni/what/vision-goals.

respective agencies involved. The NNI also continues to maintain a strong portfolio of research on ethical, legal, and other societal issues related to nanotechnology.

Objective 4.1: Incorporate safety evaluation of nanomaterials[16] into the product life cycle, foster responsible development, and where appropriate, sustainability across the nanotechnology innovation pipeline.

Individual Agency Contributions to Objective 4.1

CPSC: CPSC has been maintaining a comprehensive inventory of consumer products containing nanomaterials using proprietary business information acquired through contractor reports. The CPSC staff identifies products that claim to contain, or are believed to contain, nanomaterials, and maintains a database with detailed information on these products.

CPSC has purchased a number of contractor reports describing the commercialization of nanomaterials, but these reports were not received until late 2012. Based on these reports, the CPSC staff is working to complete updates to the internal database of products that contain nanomaterials. Staff members continue to review the reports and plan to complete a draft report summarizing the information about products on the market that contain nanomaterials in 2013. In June 2012, CPSC hosted a meeting of the International Life Sciences Institute (ILSI) Nano Release Consumer Products Project that provided an overview of the commercialization of carbon nanotubes and release scenarios from consumer products. A wide range of stakeholders attended the meeting.

EPA: EPA is evaluating the life cycle of consumer products containing nanomaterials, starting with raw material extraction and ending with ultimate disposal. This research is assembling a complete life cycle inventory for specified nanotechnology products. Using the principles of ―green chemistry" and ―life cycle assessment," EPA is developing more environmentally friendly methods to produce nanomaterials and crafting innovative technologies that minimize the impacts of engineered nanomaterials on human health and the environment. The innovative technologies developed by EPA are being used in environmental remediation technologies and commercialized in collaboration with private companies. EPA is also evaluating the use of zero-valent iron nanoparticles, which is an emerging option for the treatment of contaminated soil and groundwater, as well as cerium oxide that is being added to fuel used in diesel-powered buses to reduce emissions.

FDA: The use of nanotechnology may alter safety, effectiveness, performance, and/or quality of FDA-regulated products. Nanomaterials can present regulatory challenges because properties of a material relevant to the safety, effectiveness, performance, or quality of a product may change as the size of the material enters into, or varies within, the nanoscale range. For this reason, FDA is interested in additional scientific information and tools to help better detect and predict potential effects of such changes on human or animal health.

Continuing the course it began when FDA formally joined the NNI budget crosscut for the first time in 2011, FDA will conduct activities relevant to PCA 7 (Environment, Health, and Safety) that support the following agency-wide priorities: (1) staff training and professional development, (2) laboratory and

---

[16] Consistent with other NNI EHS documents, the term *nanomaterials* here refers to engineered nanoscale materials, i.e., those materials that have been purposely synthesized or manufactured to have a least one external dimension of approximately 1-100 nanometers—at the nanoscale—and that exhibit unique properties determined by this size. Nanotechnology-enabled products encompass intermediate products that exist during manufacture as well as final products.

product testing, and (3) collaborative and interdisciplinary research related to product characterization and safety. Together, these investments support the responsible development of nanotechnology.

NIH/NIEHS: NIH is presently invested in a number of research efforts relevant to human health, including characterization of nanopharmaceuticals and efforts to assess toxicology and human exposure.

NIEHS will invest in the following priority areas to gain knowledge on ENMs:

* The relevance of *in vitro* and *in vivo* test methods and the integration of results from these methods in building computational models to predict human health outcomes.
* The relationship between physicochemical properties of ENMs and their pharmacodynamics, body burden, and associated health risks.
* The identification of principles of ENM interaction with biological systems, protein/lipid opsonization, and the interaction of biologically modified ENMs with biological systems, and integration of this knowledge to interpret human health risks.
* The characterization of exposure and its interaction in biological matrices, and understanding the relationship of life stage and host susceptibility factors in the human responses to ENM exposure.
* The development of sensor-based personal monitoring tools for ENM exposure assessment, and integration of those tools with health effects assessment in occupational exposure.

NIOSH: NIOSH has continued its efforts to develop more complete hazard and safety assessments of key members of major classes of engineered nanomaterials: carbon nanotubes, metal oxides, silver, and the nanowire forms of silver, silica, and titanium. New research results were presented demonstrating the effectiveness of a —sfer by design" strategy by demonstrating decreases in adverse biologic activity by modifying surface functionality. NIOSH sponsored a Safe Nano Design workshop that brought together key program areas of toxicology, risk assessment, facility and process design, and EHS program management to develop guidance that promotes sustainability across the life cycles of nanomaterial-enabled products. Impact in this key program area has been realized through private sector companies partnering with NIOSH to conduct on-site research that results in the development and implementation of effective risk management practices that ultimately result in safer, more efficient nanomanufacturing processes.

NIST: NIST is developing measurement tools, essential for safety evaluation of ENMs, for determination of the physico-chemical and toxicological properties of key ENMs in relevant media (including air, water, blood, and biological tissue); detection of ENMs in such media and in nanotechnology-enabled products; and assessment of transformations of ENMs over time. In 2013 and 2014, NIST plans to develop transferable methods and standards to assess the release of ENMs from nanotechnology-enabled products, including composites, during their life cycles, to sample air containing airborne ENMs, and to determine the concentration and physico-chemical properties of airborne ENMs. Recent accomplishments and key plans include the following:

* NIST recently issued the world's first titanium dioxide nanoparticle and single-wall carbon nanotube (SWCNT) —soot" reference materials, and published five protocols for the dispersion of titanium dioxide nanoparticles in various media. In 2013 and 2014 NIST will release the world's first silver nanoparticle and length-sorted SWCNT reference materials, along with associated protocols for sample preparation and measurements.
* NIST is designing toxicology assays for ENMs that enable evaluation of nine quality metrics for describing assay performance, and in 2013 and 2014, it will identify the sources of assay variability and demonstrate robustness through pilot interlaboratory testing.

- NIST is developing methods to characterize CNT dispersion in real-world products and to measure CNT accumulation at surfaces and elucidate mechanisms that may lead to the release of CNTs into the environment as the products age and abrade.
- NIST is developing standards for electron microscopy measurements of nanomaterials supported by innovative image acquisition and sample cleaning methods.

NSF: In 2014, NSF will continue its funding for EHS research (PCA 7) at $29 million, representing about seven percent of its overall NNI budget. Requests for research are primarily directed at EHS implications and methods for reducing the respective risks of nanotechnology development.

NSF sponsored the Sustainable Nanotechnology Organization inaugural conference on 4–5 November 2012. NSF's Nanoscale Science and Engineering Grantees Conference with Focus on Environment was held on 3–4 December 2012. NSF, EPA, and USDA plan to have a similar joint conference in December 2013.

USDA/NIFA: The AFRI nanotechnology program has been and will continue supporting the EHS research targets that are most relevant to agriculture production and food applications. Its 2012–2013 combined RFA guidance states: —T◌ensure responsible development and deployment of nanotechnology and reap the benefits, applications should consider incorporating proper risk assessment studies as appropriate. These may include characterization of hazards and exposure levels, transport and fate of nanoparticles or nanomaterials in crops, soils (and soil biota), and livestock. This may also include animal feed formulations and processes that utilize novel materials or develop new nanostructured materials or nanoparticles that are bio-persistent in digestive pathways." In addition, research funded under the AFRI Physical and Molecular Mechanisms of Food Contamination program includes projects aiming to elucidate the physical and/or molecular mechanisms that allow engineered nanoparticles to attach onto and/or internalize into fresh and fresh-cut produce, including nuts (also applies to Objective 4.4).

Coordinated Activities with other Agencies and Institutions Contributing to Objective 4.1

CPSC, NIH, HHS: The Household Products Database (HPD) of the National Library of Medicine (NLM) (http://www.householdproducts.nlm.nih.gov) provides information online for thousands of consumer products. This database provides information to consumers, scientists, and other stakeholders on the chemicals contained in brand-name products and the potential health effects of these chemicals. In 2012, a modification of an MOU was signed to identify approaches to enhance the NLM database. This work will develop information that will provide consumers with knowledge about the uses of nanomaterials in products and the uses and health effects of nanoparticles.

CPSC, NIOSH: In 2008, CPSC initiated an interagency agreement (IAG) with NIOSH to evaluate the particulate aerosol generated during use of an antimicrobial spray product containing titanium dioxide (TiO_2) nanoparticles. Aerosol products containing nanoparticles have a wide variety of uses and applications, and there is concern about nanomaterial exposures during consumer use and in occupational settings and in the environment. CPSC has provided funding for the construction of a generation system and chamber to test the various aerosol products that are on the market, and NIOSH has provided the expertise and staff time for the evaluation. The project has been successful at identifying nanomaterials in products. In 2010, under another IAG between CPSC and NIOSH, NIOSH conducted testing to determine the exposure impact of antimicrobial sprays that contain nanomaterials. In 2011 and 2012, these tests will be conducted on additional products. In 2012, NIOSH completed a significant phase of this research. A manuscript has been published in the *Inhalation Toxicology Journal,* and two status reports have been completed. The first report provides the exposure and health effects data that were generated during the evaluation of nanomaterials in selected aerosol products, and the second focuses on the effects of nanosilver particles.

In 2012, a status report was completed describing the testing, conducted by NIOSH under an IAG between CPSC and NIOSH, of nanoparticles incorporated into inks and toners used in home office equipment. The report also described the concentrations and characteristics of airborne nanoparticles released during use.

CPSC, NIOSH, EPA: The unique properties of nanosilver are being used in consumer products, including room sprays, laundry detergents, wall paint, clothing textiles (such as shirts or underwear), and products intended for use by children (such as teething rings and plush toys). Exposure associated with silver varies with the chemical form (metallic, salt) and the route of exposure (ingestion, inhalation, or dermal contact).

Evaluating potential exposures to consumers from use of nanosilver-enabled products is critical for assessing potential health effects. Results obtained from available studies are highly variable, precluding generalization from these studies to other consumer products. In 2011, NIOSH, in collaboration with EPA and CPSC, conducted product testing using scientifically credible protocols to evaluate the exposure potential to nanosilver from consumer products. In 2012, the draft report was completed; CPSC anticipates publication later this year.

CPSC, NIST: A collaborative research effort between CPSC and NIST will (1) develop protocols to assess the potential release of nanoparticles into the indoor air from various consumer products, and (2) determine the potential exposure to human occupants. Measurement protocols do not exist yet to characterize these particle emissions or to assess the properties (i.e., size, shape, and composition) of the emitted particles that may relate to any health impacts.

In 2012, CPSC signed an IAG with NIST[17] to conduct testing to determine the release of nanoparticles from a range of paints and coating materials into indoor air. This project will meet an important objective in developing methods to distinguish engineered nanoparticles from those that may be produced incidentally in the home. A status report describing the measurement protocols and methods that have been developed for laboratory testing was completed in 2012.

Carbon nanotubes are reported to be incorporated into sports equipment, such as baseball bats and golf clubs. Nanotubes, cylindrical nanostructures where the length of the tube could be much greater than its nanoscale diameter, present great material strength and are also lightweight. The extent to which carbon nanotubes may be released from sports equipment during use and misuse scenarios is unknown. In 2012, CPSC signed a modification of the existing MOU[18] with NIST and began the next phase of testing products containing CNTs. A status report has been completed describing the work done under this MOU.

CPSC, NSF, NNI agencies: In 2012, CPSC signed an IAG with NSF to support research on the release of nanomaterials from clothing during consumer-use scenarios.[19] This solicitation was developed to meet the CPSC's data needs. A status report on the research conducted under this solicitation has been completed.

EPA, NSF: Under its Science to Achieve Results (STAR) program, EPA is planning to fund, in collaboration with NSF, two interdisciplinary solicitations: the first solicitation, the Networks for Sustainable Molecular Design and Synthesis (NSMDSs) is focused on facilitating the safe design strategies, processes, and pathways (including catalytic pathways) that consume less fresh water, generate less waste, and use less energy than current practice. These new approaches will minimize hazards that arise not only from chemical structure and intended use, but also from their synthesis, production, consumption, reuse, and disposal. The second solicitation, Networks for Characterizing Chemical Life Cycle (NCCLCs), is focused on

---

[17] The IAG is available at: http://www.cpsc.gov/PageFiles/93423/CPSC-I-12-0007.pdf.

[18] See http://www.cpsc.gov/PageFiles/94333/CPSC-I-12-0009.pdf for contract details.

[19] The solicitation can be found at http://www.cpsc.gov/library/foia/foia12/contracts/CPSC-I-12-0014.pdf.

examining the life cycles of synthetic chemicals and materials as they relate to their manufacture, use, transport, and disposal or recycle.

FDA, NNI agencies, international organizations: As needed and appropriate, FDA continues to foster and develop collaborative relationships with other Federal agencies through the NNI, as well as with sister regulatory agencies, international organizations, healthcare professionals, industry, consumers, and other stakeholders.

NIH/NIEHS, FDA, NIOSH: Several interagency efforts are underway in support of Objective 4.1. For example, The NIEHS Centers for Nanomaterials Health Implications Research (NCNHIR) consortium was formed in 2010. This consortium aims to develop predictive models to assess potential health effects supported by systematic understanding of ENM interactions at cellular, molecular, organ, and whole animal organization levels. Also, an interagency effort with National Center for Toxicological Research at FDA and NIOSH is working to address the potential human health hazards associated with the manufacture and use of a representative cross-section of several different classes of nanoscale materials.

NIOSH, CPSC, NIH/NIEHS/NTP: NIOSH continues formal collaborations with the CPSC and the National Toxicology Program to deliver research results that provide basic knowledge in two fronts. Work done with CPSC continues to characterize the exposure potential of selected nanomaterials from consumer products in support of human exposure risk assessments. NIOSH continues its collaboration with NTP to investigate the actual worker exposure potential during manufacture and use of engineered nanomaterials.

NIST, EPA, NIOSH, FDA, CPSC, CDC, OSHA, industry: Federal agencies are working with industry on the NanoRelease Project, coordinated by the Research Foundation of the International Life Science Institute, to identify available methods for the evaluation of the release of multi-walled CNTs from polymer-based composites, identify gaps where new methods are needed, and eventually fill gaps by testing and evaluating select methods.

NIST, FDA, EPA, NIOSH, CPSC: With input from other agencies, NIST is developing measurement methods and reference materials for nanosilver for EHS assessments of products containing this nanomaterial.

NSF, EPA, NIST, CPSC: The National Science Foundation and the U.S. Environmental Protection Agency have funded and supported the development of two university-based Centers for Environmental Implications of Nanotechnology. These centers focus on the environmental and public health implications of nanotechnology and have focused on the fate and transport of nanomaterials in the environment and exposures of various organisms. The NSF centers, in collaboration with NIST, completed a review of studies related to previous findings and experimental procedures for quantifying release of, and exposure to, nanomaterials from treated products. In addition, NSF was able to move ahead and complete much of the Phase 1 testing that was proposed in the interagency agreement. A status report on the progress of the study and plans for completing the next phase of the investigation has been completed.

USDA/FS, NIOSH: To ensure the health and safety of pilot plant workers, the FS FPL is working with NIOSH to establish occupational safety protocols at the newly commissioned cellulose nanocrystals pilot plant. The occupational safety protocol development started in 2012 and is expected to be completed in 2013. FS and NIOSH will use the FPL pilot plant experience to assist the private sector to establish occupational safety protocols in CNC production facilities.

Objective 4.2: Develop tools and procedures for domestic and international outreach and engagement to assist stakeholders in developing best practices for communicating and managing risk.

Individual Agency Contributions to Objective 4.2

EPA: EPA is researching how nanomaterials affect ecosystem and human health, evaluating methods for detecting, quantifying, and characterizing nanomaterials, as well as assessing, ranking, and predicting nanomaterial health effects using rapid, automated chemical screening called high-throughput testing methods. EPA is also evaluating exposure to nanomaterials at multiple levels and will provide qualified and credible alternative testing methods that can be used to predict toxicity. EPA is developing a database for hazard identification of structure–activity relationships and identifying unique nanomaterials properties that can be used to design or engineer safer nanomaterials. EPA is also assessing the potential effects of nanoscale titanium dioxide (used in cosmetic and sunscreen products), nanosilver, carbon nanotubes, nanocerium oxide, and nanocopper on the health of ecosystems. Researchers are testing toxicity of nanomaterials in aquatic species, determining modes of exposure, and how these materials can accumulate in both marine and freshwater (lakes, streams) ecosystems.

Coordinated Activities with other Agencies and Institutions Contributing to Objective 4.2

CPSC, EPA, UK agencies: EPA and similar agencies in the United Kingdom have collaborated in soliciting proposals for research into the potential environmental and public health impacts of nanomaterials. Internationally known experts on nanotechnology environmental, health, and safety issues served as members of the panel to select the research proposals. In 2010, CPSC staff participated in the process to select one particular research proposal to quantify exposure to consumers and to develop risk models to predict potential health effects. In 2011, the study investigators in the United States and the United Kingdom coordinated research activities, set up equipment and laboratories to conduct the studies, and synthesized the nanoparticles that will be tested in the selected consumer products.

In 2012, a status report describing the data generated by the research supported through this collaboration was completed. The project team adapted existing consumer exposure models in order to develop modules for consumer exposures and exposures in occupational settings. Application case studies done in 2012 focused on silver nanoparticles. The researchers are part of an international team whose collaboration is serving as a model for systematically addressing complex problems associated with nanoparticle risk assessment. The results will expand the current limited knowledge about health risks associated with the use of nanotechnology-based consumer products, particularly consumer sprays containing nanoparticles.

EPA, FDA, NIOSH, NIST, OECD: The OECD's Working Party on Manufactured Nanomaterials (WPMN) is aimed at assisting countries in their efforts to assess the safety of nanomaterials and has the active participation of U.S. agencies, including EPA, FDA, NIOSH, and NIST. The OECD program has focused on methods to ensure the safety of nanomaterials, including the testing of 14 commercial nanomaterials for 59 endpoints in accordance with current OECD test guidelines (the Sponsorship Programme); the development of new guidance, test guideline modifications, and alternative methods; exposure measurement and mitigation; environmental sustainability; and risk assessment and management. By 2014, several dossiers that address testing on key nanomaterials are expected to be completed; these dossiers will illustrate data trends across nanomaterials, applicability of OECD protocols, and areas for new protocol and guidance development. These areas will be explored in a series of parallel workshops in 2013 and 2014, including ones on environmental fate and ecotoxicity, physical-chemical properties, genotoxicity, and the grouping of nanomaterials. A new document —Gidance on Sample Preparation and

Dosimetry for the Safety Testing of Manufactured Nanomaterials"[20] that is applicable to regulatory testing of nanomaterials was recently published on the OECD website, and at least four OECD test guidelines are currently under consideration for modification in the 2013–2014 timeframe.

NIH (NCI, NIEHS), NIST, FDA: The 2011 NNI EHS Research Strategy provides guidance to the Federal agencies that produce the scientific information for risk management, regulatory decision-making, product use, research planning, and public outreach. NCI, NIST, and FDA continue to collaborate to develop reference materials and relevant measurement protocols for biomedical applications. NIH/NIEHS scientists played a significant role in the development, authorship, and review of the document. Additionally, supported by the NIH Common Fund's Regulatory Science program, NIH and FDA have funded a project to develop a new method to predict the potential harmful effects of nanoparticles intended for use in clinical applications prior to their testing in humans. If successful, this approach will provide a safety review for potential new therapeutic candidates.

NIOSH, industry, global partners: NIOSH will continue to enlist private sector partners to participate in field research efforts aimed at evaluating worker exposure and developing risk management guidance. Three distinct activities will continue into 2013 and 2014: initial assessments for new engineered nanomaterials, assessments for carbonaceous nanomaterials, and evaluation of containment and control strategies for nanomaterial processes. NIOSH also continues to integrate a life cycle approach for following nanomaterials during initial manufacture, incorporation into an intermediate, creation of a nanotechnology-enabled product, and end-of-life of the product. All of these activities are key elements of the nanotechnology outreach effort at NIOSH. International outreach is accomplished through collaboration with research institute counterparts in other nations. NIOSH is a technical collaborator on multiple projects in the 7th Framework Programme in the EU and maintains a variety of collaborations with institutes in the UK, Finland, Canada, Brazil, and Australia.

Domestic and international dissemination of EHS risk management practices is the objective of a NIOSH-sponsored information database, the Good Nano Guide. The guide is a global collection on a wiki platform of practices available to EHS practitioners around the world. The wiki platform allows efficient collection, review, and posting of effective practices that support responsible commercialization of nanomaterials. In 2013–2015, NIOSH will sponsor the migration of the database from Rice University to Oregon State University. In 2013, NIOSH will integrate some of the initial outputs of knowledge infrastructure -nanoinformatics" with the Good Nano Guide and will also actively seek interfaces with other U.S. government-funded systems, such as the Nanomaterial Registry.

NNI agencies through the GIN Working Group: Led by DOS, members of the NSET Global Issues in Nanotechnology Working Group, with strong support from NNCO, engage internationally though a number of mechanisms and venues. They participate in bilateral Joint Committee Meetings on Science & Technology and in multilateral organizations such as the OECD, ISO, APEC, UNEP, WHO, and FAO. For example, in March 2012, under the auspices of the U.S.-Russia Bilateral Presidential Commission, the Open World Leadership Center (an independent agency in the U.S. legislative branch) and DOS co-sponsored a group of young Russian scientists to travel to the United States for discussions on nanotechnology EHS research.

NNI Agencies, European Union: The Transatlantic Economic Council has identified nanotechnology as a priority area for increased cooperation between the European Union and the United States. In March 2011, the European Commission (EC) and the NNI organized the first US–EU Workshop on Bridging NanoEHS

---

[20] Document designation as ENV/JM/MONO(2012)40; for the complete report, go to: http://search.oecd.org/officialdocuments/displaydocumentpdf/?cote=env/jm/mono(2012)40&doclanguage=en

Research Efforts. At this meeting, Communities of Research (CoRs) were proposed as a platform for scientists to develop a shared repertoire of protocols and methods, to overcome research gaps and barriers, and to enhance their professional relationships. In 2012, three CoRs were launched at the Society of Toxicology Annual Meeting in San Francisco, CA, and an additional three were launched at the NanoSafety Cluster meeting in Grenoble, France. Progress towards CoR objectives was reviewed at the second EU–US workshop on Bridging NanoEHS Research Efforts, held in October 2012 in Helsinki, Finland (http://us-eu.org/).

NSF, EPA, USDA/NIFA: A joint grantees' conference is being planned for December 2013 to evaluate the progress of the 2010 joint solicitation program by EPA, NSF, and USDA/NIFA. The fourteen projects funded by the solicitation stipulated collaboration with Europe.

Objective 4.3: Identify and manage the ethical, legal, and societal implications (ELSI) of research leading to nanotechnology-enabled products and processes.

Individual Agency Contributions to Objective 4.3

NIH: In 2012, the NIEHS director and key staff met with the NIEHS Public Interest Partners. At the request of the partners, NIEHS staff presented a primer on nanotechnology. The membership represents diverse groups, including disease, disability, and environmental education and advocacy organizations. The group lends grassroots perspectives to the research agenda of NIEHS and serves as a key contributor to the translation of research findings for the public, policymakers, and private foundations. Additionally, NIH (specifically, NIEHS/NTP and NCI/NCL) are working to develop international documentary standards in nanotechnology through the International Organization for Standardization (ISO) as coordinated by the American National Standards Institute's U.S. Technical Advisory Group to ISO TC229. NIH also participates in other standards development activities with ASTM, IECT, and IEEE.

NSF: Two NSEC centers for "Nanotechnology in Society" will continue operation in 2014 with an increased focus on anticipatory governance and public participation.

USDA/NIFA: In the AFRI Foundational Program 2012–2013 combined solicitation, USDA/NIFA will support proposals that are intended to address assessment and analysis of the perceptions and social acceptance of nanotechnology and nanotechnology-based food or non-food products by the public and agriculture and food stakeholders, using appropriate social science tools.

Coordinated Activities with other Agencies and Institutions Contributing to Objective 4.3

NSF, CPSC: CPSC has collaborated with NSF to provide additional support for the Center for the Environmental Implications of Nanotechnology at Duke University in order to allow the center to develop innovative tools for measuring the potential health impact of nanotechnologies used in the flammability treatment of thermoplastic and thermoset-based consumer products, and to develop risk models for human exposure to materials released from these consumer products. In 2011, CPSC and NSF signed an interagency agreement (CPSC-I-ll-OOl6) in which CPSC provided $0.25 million for this one-year activity.

NSF, NIH, EPA, NASA, ONR, USDA: NSF co-sponsored with NIH, EPA, NASA, ONR, and USDA the study, "Convergence Knowledge and Technology for Societal Benefit" (final workshop on 11 December 2012 at NSF), where nanotechnology was one of the main themes. The study provided an assessment on methods of convergence of advanced knowledge and technology and their application in various science and technology areas to address major national and international issues of broad concern to society (also applies to Objectives 1.1 and 1.3).

Objective 4.4: Employ nanotechnology and sustainable best practices to protect and improve human health and the environment.

Individual Agency Contributions to Objective 4.4

DOD: The U.S. Army Center for Environmental Health Research is evaluating the potential health risks associated with nanomaterials found in Army materiel. Specific goals include identifying Army materiel that incorporate nanomaterials, performing a relative risk ranking of the identified materiel, conducting more detailed risk assessments of materiel with the highest risks, and identifying data gaps and research needs most relevant to the Army that might guide future efforts. Over 100 Army nanomaterials applications have been identified so far based on a data call to Army laboratories. Data compilation and risk evaluations being conducted by the Research Triangle Institute (RTI) are scheduled for completion in 2013. These products are being developed in cooperation with the United States Army Public Health Command and will be used to assist in the health hazard assessment process for Army materiel.

The Army Engineer Research and Development Center is also investigating a quantitative means to determine the environmental and human health effects resulting from exposure to explosives, propellants, smokes, and products containing nanomaterials and new and emerging compounds and materials produced or used in Army industrial, field, and battlefield operations, or disposed of through past activities. Methods are being explored for the design of nanomaterials and other new and emerging materials such that adverse effects on human health or the environment are minimized in their designed state and when they enter the environment where they may break down.

EPA: EPA is developing a comprehensive environmental assessment (CEA) to identify and prioritize research to support the assessments and risk management decisions in the area of nanomaterials. The CEA approach structures information in a framework to support evaluations of near-term (e.g., risk-related trade-offs) and long-term priorities (e.g., research gaps) for assessment and risk management decisions. The EPA is also developing case studies pertaining to the product life cycle, environmental transport and fate, exposure dose in receptors (i.e., humans, ecological populations, and the environment), and impacts for particular nanomaterials in specific applications. Current efforts are focused on developing a case study on multi-walled carbon nanotubes in flame-retardant coatings applied to upholstery textiles.

Coordinated Activities with other Agencies and Institutions Contributing to Objective 4.4

EPA, CPSC: EPA and CPSC have collaborated on a case study to evaluate nanoscale silver used as a disinfectant spray product. EPA and CPSC are also collaborating in a nationwide effort to conduct innovative research to assess the effects of nanocopper as a fungicidal agent.

NIH, NSF, NIST, EPA, CPSC: Established by NIEHS using American Recovery and Reinvestment Act (2009) funding, the Nano Grand Opportunity consortium carried out interlaboratory comparisons of assays using defined cell culture systems and animal models to evaluate the biological effects of a well-characterized set of ENMs. Within the framework and context of the NNI EHS strategy, NIH is working actively with other Federal agencies including NSF, NIST, EPA, and CPSC to improve measurement, characterization, and safety of nanomaterials.

4. NANOTECHNOLOGY SIGNATURE INITIATIVES

In support of the President's priorities and innovation strategy, the Federal agencies participating in the National Nanotechnology Initiative have identified focused areas of national importance that may be more rapidly advanced through enhanced interagency coordination and focused investment. The Nanotechnology Signature Initiatives (NSIs) provide a spotlight on critical areas and define the shared vision of the participating agencies for accelerating the advancement of nanoscale science and technology from research through commercialization.

Research at the nanoscale spans a diverse array of science and engineering fields and is a broadly enabling endeavor. Inherently interdisciplinary, these research and development efforts greatly benefit from coordinated planning and collaboration. By combining the expertise, capabilities, and resources of appropriate Federal agencies, the NSIs accelerate research, development, or insertion, and overcome challenges to application of nanotechnology-enabled products. Each contributing agency is committed to coordinating research to achieve the expected outcomes defined in the NSI white papers in order to avoid duplication of effort and maximize the return on the Nation's research investments. The focused goals and outcomes of each signature initiative fall within the four foundational NNI goals and across several PCAs. The NSIs enable synergy among the mission-specific efforts of the agencies and as such, while this section contributes a broad strategic overview, details are reported in Chapter 3. The NSET Subcommittee, in collaboration with OSTP, released three NSI topic areas in February 2010: solar energy, nanomanufacturing, and nanoelectronics. Two additional areas were added in calendar year 2012: sensors and informatics. The subcommittee continues to examine other potential areas of nanoscale science and technology that may benefit from similar close coordination.

Combined funding for Nanotechnology Signature Initiatives has increased annually since 2012, as shown in Table 7.

Table 7: Nanotechnology Signature Initiatives, 2012–2014 (approximate funding, dollars in millions)				
Nanotechnology Signature Initiative	Participating Agencies	2012 Actual	2013 CR	2014 Proposed
Sustainable Nanomanufacturing	DOD, DOE, EPA, IC/DNI, NASA, NIH, NIOSH, NIST, NSF, OSHA, and USDA/FS	56	72	60
Solar Energy Collection and Conversion	DOD, DOE, IC/DNI, NASA, NIST, NSF, and USDA/NIFA	88	82	102
Nanoelectronics for 2020 and Beyond	DOD, DOE, IC/DNI, NASA, NIST, and NSF	92	87	80
Nanotechnology Knowledge Infrastructure (NKI)	CPSC, DOD, EPA, FDA, NASA, NIH, NIOSH, NIST, NSF, and OSHA	2	2	23
Nanotechnology for Sensors and Sensors for Nanotechnology	CPSC, DOD/DTRA, EPA, FDA, NASA, NIH, NIOSH, NIST, NSF, and USDA/NIFA	55	65	78
TOTAL		294	308	343

Considerable effort over the past year has focused on building communities of interest among the Federal participants in each of the signature initiatives. In the coming year, focus will turn to engaging the broader community within these areas through events such as embedded technical sessions in existing conferences and workshops. These efforts aim to raise awareness and build a wider community of interest including not only the Federal participants, but also those in academia and industry engaged in research, development, and commercialization efforts, such as grantees and other stakeholders. Well-established communities of interest will enable leveraging of public and private activities to further accelerate progress and enable sustained endeavors beyond the lifetime of the signature initiatives. A summary for each of the five Nanotechnology Signature Initiatives is provided in the following few pages. Additional information and all of the NSI white papers are available at http://nano.gov/signatureinitiatives.

Nanotechnology Signature Initiative: Sustainable Nanomanufacturing— Creating the Industries of the Future

This NSI is designed to develop new technologies for the manufacture of advanced materials, devices, and systems based on nanoscale building blocks and components, and for their economical and sustainable integration into complex, large-scale systems. The nanomanufacturing NSI has two thrust areas: (1) Design of scalable and sustainable nanomaterials, components, devices, and processes; and (2) Nanomanufacturing measurement technologies. This initiative initially targets three classes of materials—high-performance structural carbon-based nanomaterials, optical metamaterials, and cellulosic nanomaterials—that have the potential for significant economic impact in multiple industry sectors.

The long-term vision for nanomanufacturing is to create flexible, —bottomup" or "top-down/bottom-up" continuous assembly methods to construct elaborate systems of complex nanodevices. Furthermore, these systems by design will reduce the overall environmental and health impacts over their full life cycles. This initiative is developing the foundation for achieving this vision by establishing sustainable industrial-scale manufacturing of functional systems with relatively limited complexity based on manufactured nanoparticles with designed properties. To progress beyond demonstrations, production must be scaled up to the required throughput and yield; the generation, manipulation, and organization of nanostructures must be accomplished in a precise, controlled, and sustainable manner; and final products must perform to specification over their expected lifetimes without the release of harmful nanomaterials. The availability of high-throughput, inline metrology to enable closed-loop process control and quality assurance is a critical prerequisite for the development of cost-effective nanomanufacturing. To this end, this initiative is focused directly on the development of inexpensive, rapid, and accurate sensing and measurement techniques.

Collaborating agencies engaged in this NSI include DOD, DOE, EPA, IC/DNI, NASA, NIH, NIOSH, NIST, NSF, OSHA, and USDA/FS. The signature initiative mechanism provides enhanced communication and facilitates coordinated planning and collaboration of activities and programs that ultimately stem from the appropriate agencies. Specific agency activities in support of this signature initiative are detailed within Chapter 3 of this supplement. Example activities are highlighted below.

Agencies contributing to the nanomanufacturing NSI are coordinating with industry and academia in an ad hoc consortium to accelerate the development of bulk carbon nanotube-based multifunctional structures and manufacturing methods. This public–private partnership has demonstrated bulk CNT material systems and processes for nanomanufacturing roll-to-roll bulk CNT yarns, sheets, and tapes for multifunctional structures. Federal agencies are actively engaged with private and academic partners regarding the technical details of materials performance, protection of environmental and human health, and the development of real-time inline measurement technologies for characterization and production control.

This collaborative effort has produced first-generation lightweight coaxial cables for application in aircraft and space systems as well as electromagnetic shielding for satellites.

Cellulosic nanomaterials have high strength, low weight, and renewable attributes that make them attractive materials for potential use in the automotive, aerospace, electronics, consumer products, and medical device industries as well as in lightweight armor or ballistic glass for the military. The Federal agencies contributing to the nanomanufacturing NSI are collaborating to advance the understanding and application of cellulosic nanomaterials. The opening of a nanocellulose pilot plant at the Forest Products Laboratory of the Forest Service is enabling access to working quantities of forest-based nanomaterials for researchers and early adopters. Agencies are also involved with a public–private partnership, Agenda 2020 Technology Alliance, to leverage activities in academia and industry.

Nanotechnology Signature Initiative: Nanotechnology for Solar Energy Collection and Conversion

Solar energy is a promising energy source that has the potential to reduce U.S. dependence on fossil fuels. New innovations and fundamental scientific breakthroughs are required, however, to accelerate the development of solar energy technologies that are economically competitive with conventional fossil fuels. Agencies participating in the NNI have identified a number of physical phenomena where nanotechnology may play a critical role in overcoming current performance barriers to substantially improve the collection and conversion of solar energy.

Nanoparticles and nanostructures have been shown to enhance the absorption of light, increase the conversion of light to electricity, and provide better thermal storage and transport. Nanostructured artificial photosynthetic systems mimicking those found in nature will be important for the conversion of solar energy into chemical fuels. A deeper theoretical understanding of conversion and storage phenomena at the nanoscale, improvements in the nanoscale characterization of electronic properties, and developments that enable economical nanomanufacturing of robust devices will be critical to exploiting the benefits of nanotechnology. Product lifetime and reliability of technologies incorporating nanotechnology must also meet or exceed the performance of conventional technologies.

The goals of the solar NSI are to enhance understanding of conversion and storage phenomena at the nanoscale, improve nanoscale characterization of electronic properties, and help enable economical nanomanufacturing of robust devices. This initiative has three thrust areas: (1) Improve photovoltaic solar electricity generation; (2) Improve solar thermal energy generation and conversion; and (3) Improve solar-to-fuel conversions.

Collaborating agencies engaged in this NSI include DOD, DOE, IC/DNI, NASA, NIST, NSF, and USDA/NIFA. The signature initiative mechanism provides enhanced communication and facilitates coordinated planning and collaboration of activities and programs that ultimately stem from the appropriate agency. Specific agency activities in support of this signature initiative are detailed within Chapter 3 of this report. Example activities are highlighted below.

Extensive efforts are underway throughout the Federal Government in the areas identified in the solar NSI whitepaper. Naturally, the Department of Energy plays a significant role with programs such as the SunShot Initiative,[21] aimed at making solar energy cost competitive with other forms of energy by the end

---

[21] More information on the SunShot initiative can be found at http://www1.eere.energy.gov/solar/sunshot/

of the decade. Although not focused solely on nanotechnology, SunShot supports substantial research and development on photovoltaics and concentrating solar power, two of the thrust areas of the Solar NSI. The BRIDGE program supports collaborative efforts providing engineers and scientists developing solar technologies with the tools and expertise of the scientific user facilities. Eligibility for this program is unrestricted and can include collaborators from academia, industry, or other Federal laboratories.

The Foundational Program to Advance Cell Efficiency combines technical and funding resources of DOE and NSF and aims to reduce the gap between efficiencies of prototype cells achieved in the laboratory and those produced on manufacturing lines. DOE and NSF also co-fund the Quantum Energy and Sustainable Solar Technologies ERC, which is focused on advancing photovoltaic science, technology, and education.

Over the past several months, communication facilitated through this NSI has highlighted mechanisms for enhanced interagency collaboration, and stronger ties have developed in the process of strengthening the Federal community of interest in this area. For instance, there is an increased awareness among the agency participants of characterization tools under development in support of the NSI goals. These interactions are in addition to the co-funding of major research centers and joint programs as reported in Chapter 3.

Nanotechnology Signature Initiative: Nanoelectronics for 2020 and Beyond

The semiconductor industry is a key driver for U.S. and global economic growth and has contributed significantly to the productivity gains experienced over the past several decades. These gains are due, in part, to the continuous miniaturization enabled by advances in materials science, microelectronics design and fabrication, and manufacturing, resulting in ever smaller, faster, and more affordable devices. As length scales of electronic devices near atomic dimensions, however, the physical limitations of current architectures are approaching, and new methods for storing and manipulating information must be found in order for miniaturization to continue. Continuing to shrink device dimensions is important to further increase processing speed, reduce device switching energy, increase system functionality, and reduce manufacturing cost per bit.

Researchers are pursuing a variety of approaches to provide the scientific bases to overcome the fundamental limitations that exist in the scaling of traditional electronics. Nanoscale science, engineering, and technology will play a significant role and will likely transform the very nature of electronics and the processes by which electronic devices are manufactured. Rapidly reinforcing domestic R&D successes in these arenas could establish a U.S. domestic manufacturing base that will dominate 21st century electronics commerce. The goal of this initiative is to accelerate the discovery and use of novel nanoscale fabrication processes and innovative concepts to produce revolutionary materials, devices, systems, and architectures to advance the field of nanoelectronics.

The nanoelectronics NSI has five thrust areas: (1) Exploring new or alternative —stae variables" for computing; (2) Merging nanophotonics with nanoelectronics; (3) Exploring carbon-based nanoelectronics; (4) Exploiting nanoscale processes and phenomena for quantum information science; and (5) Expanding the national nanoelectronics research and manufacturing infrastructure network.

Collaborating agencies engaged in this NSI include DOD, DOE, IC/DNI, NASA, NIST, and NSF. The signature initiative mechanism provides enhanced communication and facilitates coordinated planning and collaboration of activities and programs that ultimately stem from the appropriate agencies. Specific agency activities in support of this signature initiative are detailed within Chapter 3 of this report. Example activities are highlighted below.

Beyond the interagency collaboration, two of the contributing agencies are actively engaged with the Nanoelectronics Research Initiative (NRI). NRI is a program of the Semiconductor Research Corporation that supports university-centered research for the development of after-the-next-generation nanoelectronics technology. NRI directs a coordinated nanoelectronics research program centered at leading universities in partnership with Federal and state government agencies and is often used as a model of a highly leveraged industry-driven public–private partnership.

Collaboration also exists between the NSI participating agencies and the SRC Focus Center Research Program (FCRP), an effort of leading companies from the semiconductor and defense industries supporting basic research at U.S. universities. Research activities focus on leveraging complex systems integration to provide new capabilities, combined with identifying new materials and device physics, to accelerate progress for new mainstream technologies beyond today's integrated circuits.

The NSF National Nanotechnology Infrastructure Network, an integrated networked partnership of user facilities based at universities, along with the NIST Center for Nanoscale Science and Technology, the DOE Nanoscale Science Research Centers, and research and manufacturing centers funded by other contributing Federal agencies directly support the fifth thrust area of this NSI. These facilities not only provide access to state-of-the-art equipment, but also foster collaboration and scientific exchange.

Nanotechnology Signature Initiative: Nanotechnology Knowledge Infrastructure: Enabling National Leadership in Sustainable Design

Nanotechnology plays a role in solving global challenges by generating and applying new multidisciplinary knowledge of nanoscale phenomena and engineered nanoscale materials, structures, and products. The data underlying this new knowledge are vast, disconnected, and challenging to integrate into the broad scientific body of knowledge. Nanoinformatics is the science and practice of developing and implementing effective mechanisms for the nanotechnology community to collect, validate, store, share, mine, analyze, model, and apply nanotechnology information. Nanoinformatics is integrated throughout the entire nanotechnology landscape, impacting all aspects of research, development, and application. An improved nanoinformatics infrastructure is critical to the sustainability of our national nanotechnology proficiency by improving the reproducibility and distribution of experimental data as well as by promoting the development and validation of tools and models to transform data into information and applications. A focused national emphasis on nanoinformatics will provide a strong basis for the rational design of nanomaterials and products, prioritization of research, and assessment of risk throughout product life cycles and across sectors that include energy; environment, health, and safety; medicine; electronics; transportation; and national security.

The goal of the Nanotechnology Knowledge Infrastructure (NKI) NSI is to provide a community-based, solutions-oriented knowledge infrastructure to accelerate nanotechnology discovery and innovation. This initiative has four thrust areas that focus efforts on cooperative interdependent development of: (1) A diverse collaborative community; (2) An agile modeling network for multidisciplinary intellectual collaboration that effectively couples experimental basic research, modeling, and applications development; (3) A sustainable cyber-toolbox to enable effective application of models and knowledge to nanomaterials design; and (4) A robust digital nanotechnology data and information infrastructure to support effective data sharing, collaboration, and innovation across disciplines and applications.

Collaborating agencies engaged in this NSI include CPSC, DOD, EPA, FDA, NASA, NIH, NIOSH, NIST, NSF, and OSHA. The signature initiative mechanism provides enhanced communication and facilitates coordinated planning and collaboration of activities and programs that ultimately stem from the appropriate agencies. Specific agency activities in support of this signature initiative are detailed within Chapter 3 of this report. Although this initiative is relatively new, considerable steps have been made to move this effort forward, as highlighted below.

A critical aspect of sharing data is an understanding of the maturity or quality of the data. Representatives from the collaborating agencies of the NKI Signature Initiative have developed a nomenclature for communicating the maturity of data. Analogous to Technology Readiness Levels, the Data Readiness Levels (DRLs) provide a shorthand method for conveying coarse assessments of data for use in validating or calibrating models or assessing other data. Since the DRLs define a language for discussing data broadly and are not specific to nanotechnology or nanoscience, they have the potential to have an impact on other initiatives, such as the Materials Genome Initiative (MGI).

Representatives from the Federal agencies contributing to the NKI are also working to identify existing resources that may become part of the cyber-toolbox, key characteristics required for such resources, and potential gaps.

The NKI Signature Initiative and the MGI have common themes and, through coordination activities, productive and open communication exists at the leadership level and among the agency representatives. Strong collaboration and interaction also exists between the NKI and related activities including public–private efforts such as the community coalition of nanotechnology and informatics practitioners steering the Nanoinformatics Roadmap[22] and associated workshops in 2010, 2011, and 2012.

Nanotechnology Signature Initiative: Nanotechnology for Sensors and Sensors for Nanotechnology: Improving and Protecting Health, Safety, and the Environment

Nanotechnology-enabled sensors are providing new solutions in physical, chemical, and biological sensing that enable increased detection sensitivity, specificity, and multiplexing capability in portable devices for a wide variety of health, safety, and environmental assessments. Compelling drivers for the development of nanosensors include the global distribution of agricultural and manufacturing facilities, creating an urgent need for sensors that can rapidly and reliably detect and identify the source of pollutants, adulterants, pathogens, and other threat agents at any point in the supply chain. The increasing burden of chronic disease such as cancer and diabetes on the aging U.S. population requires improved sensors to identify early-stage disease and inform disease management. Although several new high-performance nanosensors have demonstrated rapid response and increased sensitivity at reduced size, translation of these devices to the commercial market is impeded by issues of reliability, reproducibility, and robustness.

Furthermore, the rise in the use of engineered nanomaterials in commercial products and industrial applications has increased the need for sensors to detect, measure, and monitor the presence of nanomaterials potentially released in diverse environments along the entire product life cycle, including research and development, manufacturing, use, and disposal or recycling. Currently, there is a very limited suite of devices capable of operating in these complex environments.

---

[22] Please refer to http://nanoinformatics.org/nanoinformatics/index.php/Nanoinformatics:Roadmap_2020.

The goals of this NSI are to support research on nanomaterial properties and development of supporting technologies that enable next-generation sensing of biological, chemical, and nanoscale materials. This interagency effort coordinates and stimulates creation of the knowledge, tools, and methods necessary to develop and test nanosensors and to track the fate of engineered nanomaterials in the body, consumer products, the workplace, and the environment.

The two thrust areas for this initiative are to: (1) Develop and promote adoption of new technologies that employ nanoscale materials and features to overcome technical barriers associated with conventional sensors; and (2) Develop methods and devices to detect and identify engineered nanomaterials across their life cycles in order to assess their potential impact on human health and the environment.

Collaborating agencies engaged in this NSI include CPSC, DOD/DTRA, EPA, FDA, NASA, NIH, NIOSH, NIST, NSF, and USDA/NIFA. The signature initiative mechanism provides enhanced communication and facilitates coordinated planning and collaboration of activities and programs that ultimately stem from the appropriate agency. Specific agency activities in support of this signature initiative are detailed within Chapter 3 of this report. Example activities are highlighted below.

Considerable effort since the launch of the sensors NSI has focused on identifying ongoing efforts within the contributing agencies and building the community of interest. Existing activities and responsibilities with respect to the life cycle outlined in the white paper are being identified to distinguish specific strengths or areas that may require greater attention to accelerate sensor development. Agency representatives are also working together to identify critical roadblocks encountered by grantees in the development of nanotechnology-enabled sensors, and to help prioritize coordinated efforts.

The sensors NSI is uniquely situated to leverage and support the activities and outcomes of the other signature initiatives. Advances in support of the nanomanufacturing NSI will benefit the ability to employ nanoscale materials and features for sensors that can be reasonably made at low cost. Furthermore, the efforts to address workplace safety and interactions of nanomaterials with nanomanufacturing processes and of finished products with the environment directly relate to the second thrust area of the sensors NSI. For example, NIOSH and NIEHS are working on the development of sensor-based personal monitoring tools for engineered nanomaterials exposure assessment and integration with health effects assessment in occupational exposure. Development of future nanoelectronics will support the systems required to support the collection, analysis, and transfer of large amounts of data, while the NKI will play a role in the manipulation and sharing of that data. As part of its coordination role, the National Nanotechnology Coordination Office of the NSET Subcommittee facilitates communication throughout the NSIs to ensure maximum benefit from related activities.

An early outcome from the sensors NSI has been increased communication among the program managers who fund research and representatives from regulatory agencies. This has led to encouraging grantees to engage regulatory representatives early in the process and to design experiments to include gathering the data ultimately required to gain approval to accelerate time to market.

5. Changes in Balance of Investments by Program Component Area

The 21st Century Nanotechnology R&D Act calls for this report to address changes in the balance of investments by NNI member agencies among the PCAs. These are summarized in this chapter, principally for those agencies that are reporting significant funding or programmatic changes for 2014.

CPSC: The focus of research is expected to continue to address environmental, health, and safety concerns that are included in PCA 7. Most of the work undertaken by CPSC is in collaboration with other Federal agencies, and although future work will continue under PCA 7, components of some studies may be classified under other program component areas by agency partners.

DHS: The Department of Homeland Security has reevaluated how its NNI investments fall within the NNI PCAs, and is now reporting investments for Nanoscale Devices and Systems (PCA 3), Nanomanufacturing (PCA 5), and Environment, Health, and Safety (PCA 7) for 2012 through 2014, as shown in Tables 3–5. (In prior years these investments were reported under PCA 4, Instrumentation Research, Metrology, and Standards.) A substantial new DHS investment in PCA 3 for homeland security applications is requested for 2014.

DOD:[23] The Department of Defense does not establish funding targets for nanotechnology, which has proven to be effectively competitive based on its contributions to meeting needs and providing opportunities for future capability. New projects are awarded on a competitive basis and the balance of investment can change as a result from historical levels and predictions. Nonetheless, based on the state of nanoscience and nanotechnology research and development, the DOD expects that its primary emphasis will remain in Fundamental Nanoscale Phenomena and Processes (PCA 1), Nanomaterials (PCA 2), and Nanoscale Devices and Systems (PCA 3) with continuing emphasis on Nanomanufacturing (PCA 5) and processes associated with producing nanostructures and novel nanomaterials. Historically, DOD's Technology Base (Budget Activity 1–3) investment in nanotechnology has been approximately 50% Basic Research, about 40% Applied Research, and about 10% Advanced Technology Development. The bulk of the Small Business Innovation Research (SBIR)/Small Business Technology Transfer (STTR) investment in nanotechnology would be categorized as functionally similar to Applied Research, although with a strong, direct commercializable product or process aspect more analogous to Advanced Technology Development. The current expectation is that these overall percentages will be continued through 2014. Because of the completion of significant applied research projects during 2012, the DOD firm fiscal plans for 2014 are lower by approximately 50% than 2012 actual spending. Ongoing project solicitations and competitions may result in greater than estimated funding, but this is not guaranteed. Appendix A provides more detailed discussion of significant changes and progress within PCAs for the Military Departments and DOD agencies.

DOE: There have been no significant changes in the balance of DOE investments among the eight PCAs.

DOT/FHWA: FHWA invested $1 million in PCA 2 (Nanomaterials) in 2012 and expects to double this investment in 2014.

---

[23] Details for Department of Defense investments are provided per the statutory requirement for DOD reporting on its nanotechnology investments (10 USC §2358). Additional details on DOD activities can be found in Appendix A.

EPA: There are no significant changes in EPA's balance of nanotechnology-related investments in 2014. This investment reflects a continued focus on providing new principles for innovative chemical designs that are more environmentally sustainable and that reduce the likelihood of unwanted toxic effects of nanomaterials.

FDA: FDA investments in Environment, Health, and Safety (PCA 7) will continue to enable the agency to address questions related to the safety, effectiveness, quality, and/or regulatory status of products that contain nanomaterials or otherwise involve the use of nanotechnology; develop models for safety and efficacy assessment; and study the behavior of nanomaterials in biological systems and their effects on human health.

FDA has developed a regulatory science program in nanotechnology to foster the responsible development of FDA-regulated products that may contain nanomaterials or otherwise involve the application of nanotechnology. The FDA program establishes tools, methods, and data to assist in regulatory decision-making while providing in-house scientific expertise and capacity that is responsive to the needs of nanotechnology-related FDA-regulated products.

The Office of the Commissioner, in partnership with the FDA Nanotechnology Task Force, facilitates communication and cooperation across the agency on nanotechnology regulatory science research, both within FDA and with national and international stakeholders. The FDA Nanotechnology Task Force provides the overall coordination of FDA's nanotechnology regulatory science research efforts in the following programmatic investment areas: (1) scientific staff development and professional training, (2) laboratory and product-testing capacity, and (3) collaborative and interdisciplinary research related to product characterization and safety.

Together these programmatic investment areas will better enable FDA to address key scientific gaps in knowledge, methods, and tools needed to make regulatory assessments of nanotechnology-based products.

NASA: NASA is evaluating concepts for potential new starts in 2014 consistent with the top technical challenges identified in the Nanotechnology Space Technology Roadmap and the review of the Space Technology Roadmaps by the National Research Council.

NIH: The National Institutes of Health currently supports more than $450 million in biomedical nanotechnology research. This amount is expected to be sustained into 2014, with most funds either dedicated to ongoing projects or to new investigator-initiated awards obtained via nanotechnology-related initiatives and programs. The five ongoing nanotechnology-specific funding announcements and the three major investment programs supported in 2013 are currently active and will continue into 2014. Eighteen of the 21 NIH institutes are participating organizations on these five funding opportunity announcements. Twenty-three additional funding opportunities across NIH (issued by 7 different NIH institutes, with 20 of the 21 institutes participating) may support nanotechnology research and development as a part of their goals and objectives; eighteen of these will continue to be open announcements in 2014. Currently, these open FOAs support 75 awards through 15 NIH Institutes. Importantly, ten of the 28 FOAs are SBIR or STTR awards, supporting technology transfer through translation and commercialization of nanotechnology bench science. It is anticipated that a significant portion of the budgets for 2014 emphasizing novel nanomaterials, devices and systems, and safety (PCAs 1–3 and 7) will continue to be in investigator-initiated projects tied to these FOAs.

There are also two new NIH programs with significant nanotechnology interest that, while they do not specifically call for nanotechnology, have topics highly conducive to nanomedical pursuits; these were released in 2012 (see discussion for Goal 1 objectives in Chapter 3 of this budget supplement document).

Fiscal support for these new programs is from redistribution of funds or program expiration. No significant change in the NIH total investment is expected.

NIOSH: The National Institute for Occupational Safety and Health tracks and reports its nanotechnology totally within PCA 7: Environment, Health, and Safety. NIOSH's research activities and results support activities in other areas, such as PCA 4 (Instrumentation, Metrology, and Standards) through participating in the development of characterization methods for EHS testing; PCA 5 (Nanomanufacturing) through disseminating Prevention through Design principles and practices that facilitate scale-up in sustainable nanomanufacturing; and PCA 8 (Education and Societal Dimensions) through conducting outreach to professionals and the general public on the implications and applications of nanotechnology. Maintaining a viable program in this area is a NIOSH priority and is vital for the responsible and rapid development and commercialization of nanotechnology-based products.

For 2014, NIOSH will shift the emphasis of its program to fill knowledge gaps in five strategic areas: (1) Increase understanding of new human health hazards and related health risks to nanomaterial workers; (2) Expand understanding of the initial hazard findings on engineered nanomaterials; (3) Continue the development and dissemination of high-impact guidance materials that inform workers, employers, regulatory agencies, and decision-makers about hazards, risks, and risk management solutions; (4) Support epidemiologic studies to understand the scope of worker exposure and potential health impact; and (5) Promote and assess national adherence to risk management guidance. Work in area 3 is in direct support of the Sustainable Nanomanufacturing Signature Initiative. The NIOSH nanotechnology research program is funded totally by internally reprogrammed funds and does not anticipate an increase in funds being redirected to this priority program. Initiating work in new areas will be accomplished by decreasing effort in others. NIOSH will continue to build partnerships with private sector nanomaterial companies and will coordinate efforts with other NNI agencies, including OSHA, NIST, CPSC, EPA, and DOD. An updated edition of the NIOSH Nanotechnology Research Strategic Plan (2013–2016) describes the new areas of emphasis.

NIST: NIST's intramural research laboratories, user facilities, and services programs work at the cutting edge of science to ensure that U.S. industry, as well as the broader science and engineering communities, have the nanoscale measurements, data, and technologies to further nanotechnology innovation and industrial competitiveness. The 2014 Budget Request continues NIST's strong support for research in nanotechnology, accounting for 18% of the NIST request for Scientific and Technical Research and Services, notably sustaining the increased investment begun in 2012 for Major Research Facilities and Instrumentation Acquisition (PCA 6), and significantly increasing the investments in Nanomanufacturing (PCA 5) and in Environment, Health, and Safety—nanoEHS" (PCA 7).

NSF: NSF has increased by $4.9 million the 2014 request to a total of $430.9 million as compared to the 2012 enacted budget. In Fundamental Nanoscale Phenomena and Processes (PCA 1), the 2014 Request includes $148.8 million, an increase of $2.5 million over the 2012 enacted budget as funds in support of this program component area are transitioned from other PCAs within the Directorates' PCA allocations to support current and projected research activity related to nanochemistry and nanobiomaterials processes.

NSF has been increasing the percentage of its nanotechnology investment in Nanomanufacturing (PCA 5) over the past few years, from 5.3% in 2008 to 9.3% in 2011, 9.5% in the 2012 actual, and around 12% in the 2014 request. The nanomanufacturing program has been institutionalized since 2002. There are several NSF activities that directly support the goals of the Sustainable Nanomanufacturing NSI. In 2011, for example, more than $11 million was invested in nine Nanoscale Interdisciplinary Research Teams (NIRTs) in response to a program solicitation on Scalable Nanomanufacturing. In 2012, the Scalable Nanomanufacturing program solicitation supported 8 NIRTs with a funding level of about $9 million. Also, three Nanosystems Engineering Research Centers were funded in 2012 with an annual budget of about $3.5 million/year per center for five years.

NSF is requesting significant 2014 contributions to the nanosensors NSI (for nanobiosensors and some for nanoEHS and monitoring health) and the Nanotechnology Knowledge Infrastructure NSI (for modeling, simulation, and informatics in nanoscale science and engineering).

USDA/ARS: ARS currently invests primarily in Nanomaterials (PCA 2) and expects to maintain that emphasis through 2014.

USDA/FS: FS plans to invest a total of $7 million in nanotechnology R&D in 2014. If the budget is approved, FS plans to invest an additional $1 million in PCA 1, Fundamental Nanoscale Phenomena and Processes, and an additional $1 million in PCA 5, Nanomanufacturing.

USDA/NIFA: NIFA will continue its balanced investment in PCAs similar to previous years. The primary emphases continue to be in PCAs 1, 2, 3, 7, and 8 (Fundamental Nanoscale Phenomena and Processes; Nanomaterials; Nanoscale Devices and Systems; Environment, Health, and Safety; and Education and Societal Dimensions). The Nanotechnology for Agricultural and Food Systems program in the Agriculture and Food Research Initiative Competitive Grants Program will continue to support science and technology R&D that addresses a broad range of the most critical challenges and the most promising opportunities facing agriculture and food systems.

6. External Reviews of the NNI

The 21st Century Nanotechnology R&D Act (Public Law 108-153) calls for periodic external reviews of the NNI by the National Research Council of the National Academies and by the National Nanotechnology Advisory Panel (NNAP). Executive order 13539 designates the President's Council of Advisors on Science and Technology (PCAST) to serve as the NNAP.

Summary of NNI Progress in Addressing 2012 PCAST Recommendations

The 2012 PCAST Report to the President and Congress on the Fourth Assessment of the NNI recognizes increased efforts by the NNI in a number of areas, including promoting commercialization and coordination with industry and the release of the 2011 NNI EHS Research Strategy. However, it also notes that additional efforts are needed in strategic planning (including support for the Nanotechnology Signature Initiatives), program management, metrics for assessing impact, and crosscutting areas of EHS research (e.g., informatics, partnerships, and instrumentation development), in order to maintain our Nation's leadership role in the commercialization of nanotechnology.

NNI agencies value additional efforts in the areas of strategic planning (including NSIs), program management, metrics, and crosscutting EHS research. In 2012, the NNI agencies launched two new NSIs, on Nanotechnology Knowledge Infrastructure (NKI), and on Nanotechnology for Sensors and Sensors for Nanotechnology, which are expected to impact multiple sectors, including crosscutting EHS research, healthcare, the defense sector, and homeland security. The NNI will be releasing an updated NNI Strategic Plan in 2013, based on extensive input from stakeholders including industry, academia, and the general public, coordinated by the Executive Office of the President.

Through the coordinated management of the NSET Subcommittee, the NNI agencies' nanotechnology programs and activities are supporting the goals and implementing the objectives outlined in the NNI Strategic Plan, as demonstrated in Chapter 3 of this report. OSTP and OMB have continued to provide centralized management of NNI funding activities with the assistance of the Emerging Technologies Interagency Policy Coordination Committee (ETIPC), which provides high-level interagency policy guidance on nanotechnology.

Regarding metrics for assessing impact, Chapter 2 of this report quantifies the NNI investment portfolio, and Chapter 3 documents the outputs as they address the Strategic Plan objectives. Furthermore, NNCO, NNI agencies, and OECD's Working Party on Nanotechnology co-organized the International Symposium and Workshop on Assessing Economic Impact of Nanotechnology, held 27–28 March 2012 in Washington DC. NNCO also commissioned the current National Research Council assessment of the NNI to focus on exploring potential measures of progress and appropriate metrics to improve the tracking of Federal investments in nanotechnology research.

NNI environment, health, and safety research investments have increased from $38 million in 2006 to $121 million in the 2014 request. FDA and CPSC have joined the NNI budget crosscut in recent years, with investments focused entirely on EHS research. Current efforts are aimed at increasing coordinated and crosscutting EHS research, in keeping with the PCAST recommendations. For example, the sensors NSI includes a thrust on sensing of nanomaterials in the environment and in biological systems. The Nanotechnology Knowledge Infrastructure NSI is designed to stimulate the development of models, simulation tools, and databases that will ultimately enable prediction and design of new nanoscale materials with desired properties, and inform risk assessment and management of nanomaterials. Both of these represent crosscutting and coordinated research that promotes accelerated progress in understanding and addressing potential EHS implications of nanotechnology. Numerous joint research solicitations, interagency agreements, and other interagency partnerships in the EHS arena are documented in Chapter 3 of this report. The NEHI Working Group provides the necessary coordination of NNI EHS research and regulatory activities at the programmatic level. The ETIPC also provides a higher-level coordination function in this area. Finally, the EHS portfolio and activities continue to be reevaluated in the framework of the NNI EHS Research Strategy, and Federal agencies plan to periodically update the research strategy to reflect activities in this area.

Other Ongoing Reviews

A new triennial review of the NNI was initiated in the fall of 2011 by the National Research Council of the National Academies, with a final report anticipated in the spring of 2013. The overall objective for this review is to make recommendations to the NSET Subcommittee that will improve the value of the NNI's strategy and portfolio for basic research, applied research, and applications of nanotechnology to advance the commercialization, manufacturing capability, national economy, and national security interests of the United States. Toward this objective, this National Research Council review of the NNI includes the tasks listed below:

- Examine the role of the NNI in maximizing opportunities to transfer selected technologies to the private sector, provide an assessment of how well the NNI is carrying out this role, and suggest new mechanisms to foster transfer of technologies and improvements to NNI operations in this area where warranted.
- Assess the suitability of current procedures and criteria for determining progress towards NNI goals, suggest definitions of success and associated metrics, and provide advice on those organizations (government or non-government) that could perform evaluations of progress.
- Review NNI's management and coordination of nanotechnology research across both civilian and military Federal agencies.

Another National Research Council report entitled a *Research Strategy for Environmental, Health, and Safety Aspects of Engineered Nanomaterials,* commissioned by the Environmental Protection Agency, was found to map well with the 2011 NNI EHS Research Strategy. Both National Research Council and NNI strategies find that:[24]

- The NNI has been effective in moving forward nanomaterials-related EHS research.
- It is important to utilize life cycle analysis to identify and assess nanomaterial EHS research needs.
- Stakeholder participation and engagement is critical to a nanomaterials EHS research strategy.
- In implementing an EHS research strategy, it is important that the strategy be reviewed regularly and adapted to evolving research needs.
- More research is needed on human exposure to nanomaterials (the NNI went further and identified four specific research needs in this area).
- More research is needed on nanomaterial effects on human health (the NNI has identified six specific research needs in this area).
- More research is needed on nanomaterial effects on the environment (the NNI has identified five specific research needs in this area).

---

[24] Refer to http://nano.gov/node/736 to see specific excerpts and paired references from the NNI and National Research Council reports.

- There is a need for better tools to measure and evaluate nanomaterials (the NNI has identified five specific research needs in this area).
- There is a need to develop an informatics infrastructure for nanotechnology-related EHS research.

APPENDIX A. SUPPLEMENTARY INFORMATION ON CHANGES BY PCA FOR THE DEPARTMENT OF DEFENSE[25]

Changes in Balance of Investments across PCAs by Military Departments and DOD Agencies

Army: The Army Research Laboratory (ARL) investment strategy and overall investment remain about constant with emphasis in nanomaterials, multifunctional fiber processing technologies, fabrication, characterization, and performance measures to enable revolutionary concepts for future force lethality and survivability. It should be noted that efforts in Nanomaterials for Vehicle Protection efforts have been shifted to Emerging Materials and Processes and to the new Collaborative Research Alliance, Materials in Extreme Dynamic Environments (MEDE), and a part of the Enterprise for Multiscale Research of Materials. The U.S. Army Engineer Research and Development Center (ERDC) investment remains about constant and focused on Fundamental Nanoscale Phenomena and Processes (PCA 1), Nanoscale Devices and Systems (PCA 3), and Nanotechnology Environmental Effects (PCA 7) but with an increasing emphasis on materials modeling applied research.

Air Force: The President's 2014 Budget Request for the Air Force rebalances support of nanotechnology activities into comparatively fewer but more focused programs. Fundamental nanoscale phenomena and processes (PCA 1) will have a slight reduction and there is a modest decrease in nanomaterials (PCA 2) due to the conclusion of the nanoenergetics program and nanostructured biological materials program. Nanoscale devices and systems (PCA 3) shows a similar decrease with the conclusion of the Thermal Sciences program. Nanomanufacturing (PCA 5) sees a dramatic decline in 2014 due to the completion of the advanced manufacturing program, which is largely due to the expiration of Congressional Interest Items. The successful activities of these various programs are recaptured and focused in follow-on activities in 2014 where both nanomaterials and nanoscale devices see an additional decline with the conclusion of the Bio-X initiative and the Applied Metamaterials program. The completion of the Nanoenergetics, Nanostructured Biological Materials, Thermal Sciences and Applied Metamaterials programs lead to a 2012 to 2014 total reduction of 16% overall in the NNI investment. These program resources were significantly reinvested into more focused research programs that build on the successful technical products of these previous efforts. Growth in the level of effort in newly formed programs on nanotransport, quantum systems, ceramics, and optical materials which will feed the investment in both nanomaterials and nanoscale devices and systems is possible with technical progress and available funding.

DARPA: The President's 2014 Budget request for DARPA will continue to support research in nanotechnology across two key PCAs: Fundamental Nanoscale Phenomena and Processes (PCA 1) and Nanoscale Devices and Systems (PCA 3). Compared with 2012 levels, the estimates for 2014 will decrease the budget for PCA 1 by 80%. The Quantum Entanglement Science and Technology (QuEST) Program and the Focus Center Research Program (FCRP) complete in 2013. Compared with 2012 levels, the requests in 2014 will decrease the budget for PCA 3 by 93%. The Thermal Management Technologies (TMT) Program will also complete in 2013 as planned. Project maturation leads to no currently projected 2014 funding for Nanomanufacturing (PCA 5).

---

[25] Details for Department of Defense investments are provided per the statutory requirement for DOD reporting on its nanotechnology investments (10 USC §2358).

Technical Details on Changes by PCA

Changes under PCA 1 (Fundamental Nanoscale Phenomena and Processes)

Army: ARL Weapons and Materials Research Directorate researchers have provided a theoretical basis for the selection of kinetically stabilizing alloying elements in nanocrystalline materials, and proved grain size stabilization in nanocrystalline metallic systems by experimental methods for better-performing ceramic armor materials. They continue to research novel composite materials that demonstrate self-healing capability using bio-engineered concepts. In addition, efforts are continuing to advance the principles of inverse material designs with a specific focus on use of emerging material models for future armor designs.

ERDC researchers have created techniques to provide measurements at the nano- to microscale for validation and verification of simulations of materials. These techniques will improve the understanding of bio-lamina creation and how or if those processes can be exploited for synthesis and self-healing. They have also constructed a comprehensive data set for the binding properties of nanoscale munitions constituents and emerging contaminants in biological/physiological networks to predict impacts to ecological receptors.

Navy: Navy-funded work has enabled the demonstration of a simple, robust method for the rapid prototyping of complex DNA nanostructures in a single melting–annealing cycle from a mix of short, nonpurified, synthetic, interlocking DNA strands, which had not been carefully adjusted in ratio. The method is faster and cheaper than previous methods for crafting complex DNA nanostructures, and it has the additional advantage of relying exclusively on synthetic DNA, which increases potential applications (e.g., *in vivo* applications). The results bring DNA nanotechnology into the rapid prototyping age, and the finding is expected to accelerate research in this field.

Navy-funded work has addressed the accuracy with which DNA nanostructures form and the precision with which they assemble molecules or nanoparticles. Specifically, multimodal characterization of a single linear DNA nanostructure displaying various site-specific molecular and nanoparticle functionalization was performed. A combination of physical and optical characterization methods was used, including electrophoresis, atomic force microscopy, transmission electron microscopy, dynamic light scattering, Förster resonance energy transfer, voltammetry, and structural modeling. The study revealed the advantages and disadvantages of each of the characterization techniques and showed that each method contributes differing amounts of information and insight about the DNA nanostructure and its underlying dynamic nature. Multimodal characterizations of more complex DNA structures, such as origami, are currently being performed.

Air Force: Work funded by the Air Force Office of Scientific Research has led a University of Pennsylvania team and colleagues to create a cheaper and cleaner catalyst for burning methane. Conventional catalysts for methane combustion are composed of metal nanoparticles, and in particular palladium (Pd), deposited on oxides such as cerium oxide (CeO_2). Tweaking that approach, the researchers instead used a method that relies on self-assembly of nanoparticles. They first built the palladium particles—just 1.8 nanometers in diameter—and then surrounded them with a protective porous shell made of cerium oxide, creating a collection of spherical structures with metallic cores. Testing the material's activity, the researchers found that their core–shell nanostructure performed 30 times better than the best methane combustion catalysts currently available, using the same amount of metal. It completely burned methane at 400 degrees C.

An international, Harvard-led team of AFOSR-funded researchers has demonstrated a new type of light beam that propagates without spreading outwards, remaining very narrow and controlled along an unprecedented distance. This "needle beam," as the team calls it, could greatly reduce signal loss for on-chip optical systems and may eventually assist the development of a new class of powerful microprocessors. The needle beam (the technical term for which is a cosine-Gauss plasmon beam) propagates in tight confinement with a nanostructured metal surface. The researchers found an ingenious way to demonstrate the theorized phenomenon. They sculpted two sets of grooves into a gold film that was plated onto the surface of a glass sheet. These tiny grooves intersected at an angle to form a metallic grating. When illuminated by a laser, the device launched two tilted, plane surface waves that interfered constructively to create the non-diffracting beam.

Field emission critically depends on material properties. However, at high current densities, space charge of the electrons limits the total amount of current, and this effect is material-independent. AFOSR is funding fundamental research exploring nanoscale features and their role in field emission and current transport of electrons into plasma systems. Results have provided new insights into how quantum mechanical and material-dependent physics are coupled to classical electromagnetism. These insights are providing new algorithms into density functional theories, for true *ab initio* models of field emission.

AFOSR has been supporting work exploring the structural modifications at the micro- and nanoscale in order to generate periodic or regular modifications in the shape of magnetic materials that will result in the accumulation of surface magnetic charges. The dipolar interaction associated with such a distribution will transduce itself in a shape anisotropy. The results show that the existence of asymmetry strongly modifies the magnetic behavior of a wire. A transition was observed between two different magnetization reversal modes, from a vortex mode in circular nanowires to a transverse domain wall in very asymmetric nanowires. In addition, it was observed that the coercivity is drastically modified as a function of diameter for small-diameter nanowires. Thus, one can design asymmetric nanowires that will reverse their magnetization by either the vortex or the transverse modes simply by properly adjusting their geometric parameters. Such control and predictive power can prove useful for future applications.

Researchers at Purdue University have shown how arrays of tiny "plasmonic nanoantennas" are able to precisely manipulate light in new ways that could make possible a range of optical innovations such as more powerful microscopes, telecommunications, and computers. The AFOSR-funded researchers used the nanoantennas to abruptly change a property of light called its phase. By abruptly changing the phase they can dramatically modify how light propagates, and that opens up the possibility of many potential applications. The new work at Purdue extends findings by researchers at Harvard University. In that work the Harvard researchers modified Snell's law, a long-held formula used to describe how light reflects and refracts, or bends, while passing from one material into another. Until now, Snell's law has implied that when light passes from one material to another there are no abrupt phase changes along the interface between the materials. Harvard researchers, however, conducted experiments showing that the phase of light and the propagation direction can be changed dramatically by using new types of structures called metamaterials, which in this case were based on an array of antennas. The Purdue researchers took the work a step further, creating arrays of nanoantennas and changing the phase and propagation direction of light over a broad range of near-infrared light.

The Air Force Research Laboratory in Dayton, Ohio, has experimentally confirmed theoretical work at Rice University that foretold a pair of interesting properties about nanotube growth: that the chirality of a nanotube controls the speed of its growth, and that so-called —armchair" nanotubes (referring to the shape of the chiral angle of the carbon ring) should grow the fastest. The work is a sure step toward defining the mysteries inherent in what is called the "DNA code of nanotubes," the parameters that determine their chirality—or angle of growth—and thus their electrical, optical, and mechanical properties. Developing the ability to grow batches of nanotubes with specific characteristics is a critical goal of nanoscale research.

DARPA: The goal of the Zeno-based OptoElectronics (ZOE) program is to enable a new family of all-optical devices with ultralow energy dissipation (10–100 attojoules per operation), approximately a 100-fold reduction in the energy dissipated from the best devices available today. Zeno-based devices may form the basis of integrated optical circuits, radically new fiberoptic communications components, and as this effect scales all the way to the single-photon level, quantum communications devices as well. Approaches include microresonators, nanofibers in rubidium vapor, quantum dots on photonic cavities, and designer organic molecules optimized for two-photon absorption. The program just concluded after two phases and has demonstrated the world's lowest energy dissipation optical switch.

The Program in Ultrafast Laser Science and Engineering (PULSE) is developing tabletop X-ray imaging sources with nanometer-scale resolution. High-resolution metrology and imaging applications are critical to support extreme ultraviolet (EUV) lithography. Additionally, soft X-ray sources have the potential to image and explore the dynamics of living tissue at the subcellular level. Already, synchrotrons have demonstrated the utility of spatially coherent EUV and soft X-ray sources in many fields including physics, chemistry, materials science, and biology. However, the utility of synchrotrons is limited due to the size, cost, and availability of these large user facilities. PULSE will develop tabletop, laser-driven, coherent X-ray sources with comparable in-band flux to synchrotrons that can be operated in a laboratory or industrial environment.

The Information in a Photon (InPho) program aims to develop optical communication and imaging systems that fully exploit every received photon (individual light particles). Researchers working on the communication thrust recently demonstrated a communication system that achieved 13 bits of information for every photon incident on the receiver. This demonstration was made possible by advances in nanowire-based superconducting detectors that have shown record sensitivity and high bandwidth. Future demonstrations seek to exhibit high photon efficiency (measured in bits per photon) and, simultaneously, high communication rate (measured in bits per second) using spatial multiplexing.

The parallel goals of the MesoDynamical Architecture (Meso) program are to provide DOD with unrivaled communication, sensing, and computational capabilities by exploiting mesoscale characteristics while establishing well-defined problems to accelerate the transition to quantum engineering. Potential applications could include information transduction; efficiently moving between sound, electricity, and light on demand, and exploiting nonlinearities and fluctuations to create oscillators.

DTRA: The DTRA Basic and Applied Sciences program supports research into nanoscale phenomenology and its impact upon entire systems for sensing and recognition of radiation from nuclear materials. Work conducted at The Pennsylvania State University includes interrogation of graphene-oxide nanocomposites to provide an understanding of the interactions and related phenomena between gamma and neutron radiation and the nanoscale material graphene oxide. The work supports the development of flexible architectures that may provide a unique platform for novel radiation sensors.

CBDP: Work funded by the Chemical and Biological Defense Program includes the design and synthesis of plasmonic nanostructures that exhibit strong predictable coherent effects for spectroscopic sensing of compounds. Researchers at Rice University have added to the fundamental understanding of the underlying phenomena of such nanostructures, which is necessary in order to selectively tailor plasmonic materials for highly selective chemical or biological threat sensing. Recent work on coupled nanoscale metal disks showed optical interference effects that are extremely sensitive to changes in the local environment, such as the presence of a molecule. The researchers demonstrated precise control of this effect by controlling the metal cluster geometry. Such basic research improves the understanding of complex plasmonic phenomena, which can ultimately be exploited for a wide range of applications such as ultrasensitive chemical and biological sensing, nonlinear optics, and energy harvesting.

Changes under PCA 2 (Nanomaterials)

Army: ARL Weapons and Materials Research Directorate researchers, in collaboration with the MIT Institute for Soldier Nanotechnologies (ISN), have investigated the incorporation of nanoparticles, nanotubes, and nanofibers into materials systems to produce novel sensing capabilities for enhanced situational awareness. They continue to design novel sensors and imaging devices based on carbon nanotube, quantum dot, and photonic crystal technologies, as well as scaled-up nanometallic aluminum alloy processing to characterize performance as potential ballistic protective materials.

ARL researchers have validated a nanograined metallic structures fabrication process using thermodynamic techniques. This also includes demonstrating an enhancement to the ballistic capability of transparent materials reinforced with natural cellulose nanofibers. They will continue to design synthetic, strain-rate-dependent polymers to mimic human body tissue; design and evaluate blast-resistant cellular topologies using bio-inspired computational algorithms; demonstrate transparent, nano-architectured cellulose-based composite materials; and investigate nanoscale tungsten materials to evaluate engineering properties for ballistic launch survivability.

Continuing efforts to develop lightweight aluminum (Al) alloy materials at the ISN have been significantly advanced through work funded by an ARL/Army Research Office (ARO) single-investigator grant to MIT that has led to the development of an unprecedented and general theoretical framework to predict and design thermodynamically stable, nanostructured alloys. Most nanocrystalline materials of interest rapidly evolve to coarse-grained structures, even at low temperatures, because they tend to form in a state that is far from thermodynamic equilibrium. This in turn impedes their use at elevated temperatures and precludes the development of scalable processing methods. Recent computational and experimental efforts to circumvent this problem have focused upon the addition of alloying elements to pure material nanostructures so that they occupy grain boundary sites and minimize the energy penalty of interfaces between nanocrystals. MIT/ISN researchers developed a model that describes a free energy surface in composition-grain size space, where the mixing free energy of a nanostructured binary alloy is a function of energetic interactions both within and between grains. Using this model, the researchers identified a candidate tungsten alloy with 20 atomic percent titanium that exhibited substantially enhanced thermal stability. Preliminary efforts are underway to find nanocrystalline aluminum alloy systems that are stable against grain growth.

Navy: Nanostructured metals exhibit unprecedented strength but are notoriously brittle because dislocations, which facilitate plastic deformation, become unstable in the interiors of nanograins. New nanometals have been developed that are capable of trapping and accumulating these dislocations, resulting in metals with a combination of ultrahigh strength and good ductility (damage tolerance). This is accomplished by creating materials with microstructures containing multiple length scales and, sometimes, hierarchical structures. Two examples are a 7075 aluminum alloy with over 1 GPa ultimate tensile strength (UTS) and 10% elongation to failure, and a nickel alloy with 1.7 GPa UTS and 9% elongation. These values greatly exceed any material with similar composition ever developed.

Air Force: The recently awarded Nanoscience and Technology–Nanoenergetics Initiative began scale-up of aluminized fluorinated acrylate (AlFA), a reactive material for insensitive munitions liner applications. The AlFA material has shown promise in small-scale reactive liner testing. This 12-month scale-up activity will deliver a number of test articles to develop a database of application-relevant properties and performance, and mature the technology and its manufacturing readiness.

AFOSR-funded chemists at Rice University have successfully grown forests of carbon nanotubes that rise quickly from sheets of graphene to astounding lengths of up to 120 micrometers. The hybrid material combines two-dimensional graphene, which is a sheet of carbon one atom thick, and nanotubes into a seamless three-dimensional structure. The bonds between them are covalent, which means adjacent carbon atoms share electrons in a highly stable configuration. By growing graphene on metal (in this case, copper) and then growing nanotubes from the graphene, the electrical contact between the nanotubes and the metal electrode is ohmic. That means electrons see no difference, because it's all one seamless material. The hybrid may be the best electrode interface material possible for many energy storage and electronics applications.

One of the AFOSR material portfolios is focused on getting new properties from organic-based materials. Funded efforts use nanotechnology-related structures to create or enhance properties of organic materials, including some that use chemistry, to better prepare desirable nanostructures. Hydrophilic but oleophilic films have been created based on formation of nanostructures on surfaces using polyhedral oligomeric silsesquioxane (POSS) nanoparticles. With this combination of films acting as a separation membrane, a simple oil/water separation from their emulsion has been demonstrated using only gravity.

A simple and efficient method developed by Northwestern University and Korean collaborators has been developed for the synthesis of carbon-nanotube-bridged nanowires by assembling carbon nanotubes as segments of metallic nanowires. DNA sensors based on this architecture exhibit lower detection limits and faster response times than those based on planar carbon nanotube networks. This technique will enable further incorporation of carbon-nanotube-based architectures into various nanoelectronic and chemical and biological sensor platforms due to their ease of synthesis, high yield, and monodispersity.

An AFOSR-funded team at the University of Texas at Dallas and collaborators at Hanyang University in Korea have developed a new technology of biscrolling that enables the spinning of —unspinnable" powders and nanofibers into sewable, weavable, knotable, and washable multifunctional yarns. Biscrolled yarns can be used for batteries, fuel cells, supercapacitors, interconnects, strain sensors, artificial muscles, etc. New artificial muscles made of strong, tough, highly flexible yarns of carbon nanotubes can provide a thousand times higher rotation per length than previous artificial muscles based on ferroelectrics, shape-memory alloys, or conducting organic polymers. They twist like the trunk of an elephant and accelerate a 2,000 times heavier paddle up to 600 revolutions per minute, and then reverse this rotation when the applied voltage is changed.

An aqueous solution method has been developed to permit the growth of nanotube-reinforced aluminum (Al)-doped zinc oxide (ZnO), known as AZO, that is scalable in area and in thickness. A precision atomic layer deposition method was used to develop a heterolayer AZO structure in which individual atomic layers of Al_2O_3 were inserted into ZnO. The dispersion of nanotubes into AZO reinforces this system not only mechanically but also electrically. This new material platform was found to have ultrahigh electrical conductivity, remarkable electromagnetic interference shielding properties, low thermal conductivity, and potential for thermal energy harvesting (work conducted at Brown University and Seoul National University).

The first-ever synthesis of new, geometrically well-defined C_{70} cubes has been accomplished based on a simple solution phase reaction. It was found that mesitylene plays a critical role to define the cube shape, where a mixture of mesitylene (a —good" solvent for C_{70}) and isopropanol (a "poor" solvent) creates microreactors with rapidly increasing concentrations of C_{70}. The C_{70} cubes exhibit bright photoluminescence, which suggests exciting future opportunities to develop various fullerene-based electronic, optoelectronic, and solar energy conversion devices. The AFOSR-funded Cornell team (collaborating with Pohang University of Science & Technology in Korea) also has developed a new powerful Raman imaging method that can be applied for all carbon-based nanostructures, including C_{70} cubes, carbon nanotubes, and graphitic materials. This work led to a new Korea–U.S. industrial collaboration that includes Samsung Techwin as the industrial partner, which is in the process of incorporating the Raman imaging technique into a large-scale graphene manufacturing process.

Researchers at Rice University have created a tiny coaxial cable that is about a thousand times smaller than a human hair and has higher capacitance than previously reported microcapacitors. The nanocable was produced with techniques pioneered in the nascent graphene research field and could be used to build next-generation energy-storage systems. It could also find use in wiring up components of lab-on-a-chip processors. Its discovery is owed partly to chance, fueled by the curiosity of what would happen electrically and mechanically if one took small copper wires known as interconnects and covered them with a thin layer of carbon. The coax nanocable is about 100 nanometers wide, and the capacitance of the nanocable is at least 10 times greater than what would be predicted with classical electrostatics. It is believed that the increase is most likely due to quantum effects that arise because of the small size of the cable.

DTRA: The DTRA Basic and Applied Sciences program supports investigation of nanoscale phenomenology relevant to development of new materials, including those with prospective dual use for both the DOD and civilian medical communities. The recent publication "Synergistic Administration of Photothermal Therapy and Chemotherapy to Cancer Cells using Polypeptide-based Degradable Plasmonic Matrices" was discussed under Objective 2.2. Other examples include research for design of novel photodetectors. Researchers at the University of Nebraska are developing solution-processed ultraviolet photodetectors with a nanocomposite active layer composed of ZnO nanoparticles blended with semiconducting polymers that can significantly outperform inorganic photodetectors. The device provides improved detectivity at room temperature, two to three orders of magnitude higher than that of existing inorganic semiconductor ultraviolet photodetectors.

CBDP: Efforts funded by the CBDP are improving capabilities for sensing chemical and biological threats in complex environments. Work at Yale University on silicon nanowire biosensors for label-free detection and diagnostics of biological threats and disease biomarkers has also demonstrated utility in measuring protein binding affinities and kinetics. Real-time monitoring of protein–receptor pair interactions show promise to surpass surface plasmon resonance measurement, the current standard method for such analysis, through greater sensitivity and multiplexed high-throughput analysis useful for diagnostics and drug discovery. In general, nanoscale transduction mechanisms and related catalytic platforms rely on control of selective transport at the sub-micrometer and nanometer scales. University of California–San Diego researchers have developed self-propelled artificial nanomotors that offer selective isolation of biological targets from physiological fluids by capturing and transporting them to a clean environment. The team has reported various molecular and biological recognition strategies for these motors, with potential for increasing the sensitivity and selectivity of fluidic sensors and threat scavengers, with reduced energy requirement.

Changes under PCA 3 (Nanoscale Devices and Systems)

Army: ERDC researchers have developed foundational knowledge of nanoscale physical, chemical, and mechanical properties of materials for improved performance through computational modeling and laboratory experimental research with a focus on composite and bio-inspired materials that have exceptional properties such as tensile strength and resistance to cracking and penetration. The objective is to investigate and leverage physics-based computational models and laboratory experiments to understand the relationships between the chemical and nanostructural composition of material and performance characteristics used in protecting facilities. ERDC researchers are also creating an initial integrated modeling capability for the investigation, design, and advancement of experimental materials and properties for achievement of improved strength and durability at the nanocomposite scale (1 to 100 nm).

Navy: Navy-supported work has resulted in the development of semiconductor quantum-dot-based nanosensors for the detection of nitro-containing explosives. Specifically, CdSe/ZnS core–shell quantum dots functionalized with electron-donating ligands that bind nitro-containing explosives were used as fluorescent probes for the specific analysis of trinitrotoluene (TNT) or trinitrotriazine (RDX). Binding of the nitro-containing explosives to the quantum dot surface via electron donor–acceptor interactions leads to the quenching of the luminescence of the quantum dots and the readout signal for the sensing events. The nanosensor has enabled the sensing of TNT with a detection limit corresponding to 1×10^{-10} M, and RDX with a detection limit 5×10^{-9} M. Multiplexed analysis of different explosives can be achieved by the application of different-sized quantum dots.

Air Force: Air Force Research Laboratory nano-optics researchers recently demonstrated the use of nano-antenna materials for identification friend or foe (IFF) applications. In collaboration with Alphamicron, Inc., switchable passive emitter reflectors were demonstrated that could be viewed in conventional night vision goggles. This technology has potential, through optical antenna design, to be scaled to any wavelength band. It is important to point out that this is a passive device as well, which requires no illumination source. As a compliment to this, optical antennas have been designed and demonstrated at the single-pixel level for the specific detection of emitted polarization encoded light in any wavelength band. Further application of these techniques is planned for examining their impact in hyperspectral imaging, with demonstrations planned for 2013.

Results from the California Institute of Technology (CalTech) show the utility of plasmonic structures for CMOS photonics and the control of spontaneous emission. This is demonstrated through low-insertion-loss waveguides and large modulation of refractive index in plasmonic materials like conducting oxides, which have potential for ultracompact switching. The key feature to enable successful low insertion loss is to achieve good matching of the optical modes of the silicon-on-insulator waveguide and the plasmon waveguide. The nanoscale comes into play such that if the device thickness could be significantly reduced, say to 10 nm, the overall waveguide mode index change could approach a desired Delta ~ 1.

An optoplasmonic circuit has been developed at Harvard University, in collaboration with Korean researchers, as a new nanoscale detector for propagating surface plasmons that is sensitive enough to detect single-plasmon sources. The surface plasmons, generated by a single photon excitation and guided by silver nanowires, can be converted into electron–hole pairs in the semiconductor germanium nanowire, generating a current signal. This AFOSR-funded work suggests the exciting possibility that this on-chip detector could be coupled with other quantum optoplasmonic elements based on individual emitters such as atoms, molecules, and quantum dots.

The world's smallest three-dimensional optical cavities with the potential to generate the world's most intense nanolaser beams have been created by a scientific team led by researchers at DOE's Lawrence Berkeley National Laboratory and the University of California–Berkeley, the latter funded by AFOSR. In addition to nanolasers, these unique optical cavities, with their extraordinary electromagnetic properties, should be applicable to a broad range of other technologies, including light-emitting diodes, optical sensing, nonlinear optics, quantum optics, and photonic integrated circuits. By alternating superthin multiple layers of silver and germanium, the researchers fabricated an ―indefinite metamaterial" from which they created their 3D optical cavities. In natural materials, light behaves the same, no matter from what direction it propagates. In indefinite metamaterials, light can actually be bent backwards in some directions, a property known as negative refraction. The use of this indefinite metamaterial has enabled the scaling down of the 3D optical cavities to extremely deep subwavelength (nanometer) size, resulting in a ―hyperboloid iso-frequency contour" of light wave vectors (a measure of magnitude and direction) that has supported the highest optical refractive indices ever reported.

A team of AFRL-funded researchers at Stanford's School of Engineering has demonstrated an ultrafast nanoscale light-emitting diode that transmits data at 10 billion bits per second while using thousands of times less energy than current technologies. This nanophotonic, single-mode LED can potentially perform the same tasks as a laser but is 2,000 times more energy efficient than current devices. Low-power electrically controlled light sources are vital for next-generation optical systems.

AFOSR-funded work at University of California–Los Angeles is demonstrating that the combination of 3D engineered heterostructures and high-Q cavities results in low-threshold, room-temperature, continuous wave lasing. With a threshold power density of 75 W/cm^2, this represents the first nanowire- or nanopillar-based laser with performance comparable to planar grown devices. Other results on nanolasers and nanosources include an electrically injected room-temperature lead selenide quantum-dot laser on (001) silicon, and ultralow-threshold polariton lasing from a single gallium nitride nanowire achieved by research at the University of Michigan.

A nanogenerator has been successfully developed to prove sustainable power for driving a sensor. Using the energy harvested from mechanical vibration, it was demonstrated that the nanogenerator can produce enough energy to not only drive the operation of the sensor but also allow wireless transmission of the signal for a distance of more than 10 m. This could enable sensor networks, a key technology for healthcare, environmental monitoring, infrastructure monitoring, and more. This collaborative U.S.-Korea project was funded by AFOSR.

Arizona State University researchers funded by AFOSR and ARO are finding ways to improve infrared photodetector technology that is critical to national defense and security systems. A significant advance has been the discovery of how infrared photodetection can be done more effectively by using certain materials arranged in specific patterns in atomic-scale structures. Crystals are formed in each layer; these layered structures are then combined to form superlattices. Photodetectors made of different crystals absorb different wavelengths of light and convert them into electrical signals. The conversion efficiency achieved by these crystals determines a photodetector's sensitivity and the quality of detection it provides. The unique property of the superlattices is that their detection wavelengths can be broadly tuned by changing the design and composition of the layered structures. The precise arrangements of the nanoscale materials in superlattice structures help to enhance the sensitivity of infrared detectors. The research team is using a combination of indium arsenide and indium arsenide antimonide to build the superlattice structures. The combination allows devices to generate photo electrons, which are necessary to provide infrared signal detection and imaging.

DTRA: Research funded by the Basic and Applied Sciences Directorate is contributing to the development of novel indicators for ionizing radiation. For example, Oak Ridge National Laboratory (ORNL) is working to develop nanoparticle solar blind scintillators for remote radiation detection. Nanoscintillators will enhance luminescent properties for scintillation-based devices, allowing more transparent, more effective coverage. In addition, these novel scintillation devices are expected to facilitate high signal contrast that supports remote detection under all environmental conditions by working in low solar background windows. ORNL is in the process of transitioning the materials into a contamination detection approach for detection of Technetium-99 (^{99}Tc) contamination in DOE manufacturing facilities slated for decommissioning.

Changes under PCA 4 (Instrumentation Research, Metrology, & Standards for Nanotechnology)

Air Force: AFOSR has sponsored research at the University of Illinois at Urbana–Champaign to develop a new kind of electrothermal nanoprobe to measure nanometer-scale temperature. Atomic force microscope cantilever tips with integrated heaters are widely used to characterize polymer films in electronics and optical devices, pharmaceuticals, paints, and coatings. These heated tips are also used in research labs to explore new ideas in nanolithography and data storage, and to study fundamentals of nanometer-scale heat flow. Until now, however, no one has used a heated nanotip for electronic measurements. The electrothermal nanoprobe can independently control voltage and temperature at a nanometer-scale point contact. It can also measure the temperature-dependent voltage at a nanometer-scale point contact. The electrothermal probes are different from previously demonstrated thermal nanoprobes. They have three electrical paths to the cantilever tip. Two of the paths carry heating current, while the third allows the nanometer-scale electrical measurement. The two electrical paths are separated by a diode junction fabricated into the tip. Although the cantilever design is complex, the probes can be used in any atomic force microscope. Immediate plans are to use the probes to measure the nanometer-scale properties of materials such as semiconductors, thermoelectrics, and ferroelectrics.

Changes under PCA 5 (Nanomanufacturing)

Navy: An Office of Naval Research project at SRI International is studying the development of a high-throughput nanomanufacturing platform involving massively parallel processes carried out by a swarm of microrobots. The approach is to drive microrobots with end-effectors levitated on a printed circuit board with better than 100 nm precision and repeatability. The goal is to be able to produce heterogeneous materials towards new devices and structures not achievable by current means.

An ONR-supported Multidisciplinary University Research Initiative project is developing nanoscience-based, high-speed continuous fabrication of full-function hybrid flexible electronic systems. The challenge is to form a family of devices (sensors, logic, RFID, displays, power, etc.) as a multifunctional system-on-film. The MURI is developing appropriate nanomaterials (oxides, CNTs, silver, etc.) that will be used in the building of multilayered devices; appropriate nanoscale processes (printing, imprinting, etc.) for creating the circuits in an optimal sequence, design, and construction for selected systems of interest that are optimized for performance and production; and strategies for continuous, roll-to-roll nanomanufacturing of these complex electronic systems. Flexible electronics are lightweight and low-cost, both beneficial to the military. The MURI is targeting an efficient, conformal brain–machine interface (BMI) electronic system, which will replace current discrete electrode-based BMIs.

Research within the Young Investigator Program (YIP) is developing wafer-scale nanomanufacturing technologies for plasmonic devices and flexible metamaterials to enable new generations of military applications. The project involves developing a template-stripping technique that will produce atomically smooth nanopatterned structures (holes, pyramids, etc.) in metallic films. Manufacturing challenges are structure design, accuracy, precision, and reproducibility. Potential military applications are miniaturized biochemical sensors, plasmonic photovoltaics, super-resolution imaging, infrared thermal emitters, ultrahigh-density data storage, and frequency-tunable metamaterials.

Research in another YIP project is developing scalable processing methods for fabricating vertical arrays of nanowires in thin polymer films. This is being advanced specifically in the context of high-efficiency polymer-based solar cells. The approach is to utilize magnetic fields to align nanowires such as ZnO embedded in or coated by a semiconducting polymer, which is electro-optically active as a photovoltaic material. The project will explore large-scale directed assembly of arrays for applications such as photovoltaics (PVs). Besides low-cost PVs, the technology can be used to manufacture aligned nanocomposites for applications such as desalination, energy scavenging, and metamaterials, all of which are of interest to the military.

The Defense University Research Instrumentation Program is supporting acquisition of a high-precision materials printer with high levels of repeatability, accuracy, precision, and resolution. These features will enable the development of high-performance circuits and devices in the development of nanotechnology-enabled flexible electronic systems. Such a printer will be adapted to continuous roll-to-roll mass production and other web-based nanoprinting methods.

Air Force: The CuantumFuse™ program performed by Lockheed Martin has shown several significant manufacturing demonstrations for nanocopper-based lead solder replacement. The process uses a fundamentally new approach employing nanosized copper particles and takes advantage of size-dependent melting point depression to make a fusible solder replacement that is compatible as a drop-in replacement for lead-based solder. Challenges of controlling particle size, surface properties, environmental stability, and process compatibility have been successfully addressed. This material has been successfully used in the fabrication of a test electronics board to demonstrate functionality, which shows substantial overall progress toward a lead-free solder system. The nanocopper-based solder has potential for applications in electro-optic, communications, radar, and guidance systems for air and space technology.

Researchers in Korea and at Case Western Reserve University have developed a low-cost, easier way to mass-produce better graphene sheets than the current, widely used method of acid oxidation, which requires the tedious application of toxic chemicals. Researchers placed graphite and frozen carbon dioxide in a ball miller, which is a canister filled with stainless steel balls. The canister was turned for two days; the mechanical force produced flakes of graphite with edges essentially opened up to chemical interaction by carboxylic acid formed during the milling. The carboxylated edges make the graphite soluble in a class of solvents called protic solvents, which include water and methanol, and another class called polar aprotic solvents, which includes dimethyl sulfoxide. Once dispersed in a solvent, the flakes separate into graphene nanosheets of five or fewer layers. To form large-area graphene nanosheet films, a solution of solvent and the edge-carboxylated graphene nanosheets was cast on silicon wafers 3.5 centimeters by 5 centimeters, and heated to 900 degrees Celsius. Again, the heat decarboxylated the edges, which then bonded with edges of neighboring pieces. The researchers say this process is limited only by the size of the wafer. The electrical conductivity of the resultant large-area films, even at a high optical transmittance, was still much higher than that of their counterparts formed from acid oxidation. The work was supported under the U.S.–Korea Nano-Bio-Information Technology (NBIT) Symbiosis Program through the National Research Foundation of Korea and the U.S. Air Force Office of Scientific Research.

SI2 Technologies, Inc., (SI2) has teamed with the University of Illinois Urbana–Champaign (UIUC) to leverage its experience in direct-write roll-to-roll inkjet and electrohydrodynamic (E-jet) subwavelength printing to deposit micrometer-to-millimeter conductive, dielectric, and electro-optic polymer chromophore patterned films on flexible Kapton™ and silicon (Si)-nanomembrane substrates. During the project the partners successfully demonstrated the capability of printing micrometer-scale geometries on flex substrates with conductivities approaching those of bulk metal. An Si-nanomembrane transistor and polymer waveguide was electrohydrodynamic-jet (E-jet)-printed with electrodes, cladding, and electro-optic chromophore, enabling an all-lithographic free manufacturing process to eliminate harsh etchant chemicals for improved electro-optic device performance. A novel print and plate manufacturing method was also developed for manufacturing high-power circuitry thereby mitigating skin depth effects. The potential for nanometer-scale printing continues to be explored.

Sinoora, Inc., in collaboration with Georgia Institute of Technology, is developing a low-cost, high-yield, and high-precision fabrication process for manufacturing of densely integrated photonic systems. Sinoora's mission is to address two important challenges that hinder the commercialization of functional integrated photonic systems: (1) high sensitivity of the integrated photonic devices to fabrication variations, and (2) lack of a high-throughput and low-cost manufacturing process for integrated photonic devices. Sinoora's approach is based on combining a low-cost, high-precision fabrication process using nanoimprint lithography with a novel technique for post-fabrication trimming. The unique features of

Sinoora's approach are: (1) low-cost wafer-level fabrication, (2) high yield, (3) high resolution (sub-100-nm feature sizes), (4) low device variation evidenced by high repeatability from run to run, and (5) passive trimming (or correction) of individual photonic devices (no active tuning is needed). Preliminary results indicate that the company's process will have a significant impact on the ability to scale-up the integration level of photonic integrated systems and will considerably reduce the time and cost of research and development in this field at all academic, DOD, and industry levels.

Changes under PCA 7 (Environment, Health, and Safety)

Army: A pioneering study to gauge the toxicity of quantum dots in primates has found the tiny crystals to be safe over a one-year period. The research is likely the first to test the safety of quantum dots in primates. Scientists found that four rhesus monkeys injected with cadmium-selenide quantum dots remained in normal health over 90 days. Blood and biochemical markers stayed in typical ranges, and major organs developed no abnormalities. The animals didn't lose weight. Two monkeys that were observed for an additional year also showed no signs of illness. Quantum dots are tiny luminescent crystals that glow brightly in different colors. Medical researchers are eyeing the crystals for use in image-guided surgery, light-activated therapies, and sensitive diagnostic tests. The new toxicity study—completed by the University at Buffalo, the Chinese People's Liberation Army General Hospital, China's Chang Chun University of Science and Technology, and Singapore's Nanyang Technological University—begins to address the concern of health professionals who worry that quantum dots may be dangerous to humans. The new toxicity study was supported by the John R. Oishei Foundation, Air Force Office of Scientific Research, Singapore Ministry of Education, Nanyang Technological University, the Beijing Natural Science Foundation, and the National Natural Science Foundation of China.

APPENDIX B. GLOSSARY

Act	Public Law 108-153, the 21st Century Nanotechnology Research and Development Act, 15 USC 7501
AFOSR	Air Force Office of Scientific Research
AFRI	Agriculture and Food Research Initiative (USDA/NIFA)
AFRL	Air Force Research Laboratory
Agencies	Departments, agencies, and commissions within the Executive Branch of U.S. Federal Government
ARL	Army Research Laboratory
ARO	Army Research Office (DOD)
ARS	Agricultural Research Service (USDA)
BIS	Bureau of Industry and Security (DOC)
BRIDGE	The Bridging Research Interactions through collaborative Development Grants in Energy
CBDP	Chemical and Biological Defense Program
CDC	Centers for Disease Control and Prevention (DHHS)
CEIN	Center for Environmental Implications of Nanomaterials (UCLA)
CEINT	Center for the Environmental Implications of NanoTechnology (Duke University)
CMOS	complementary metal-oxide semiconductor
CNST	Center for Nanoscale Science and Technology (DOC/NIST)
CNT	carbon nanotube
CoR	Communities of Research
CPSC	Consumer Product Safety Commission
CR	Continuing resolution
DARPA	Defense Advanced Research Projects Agency
DHHS	Department of Health and Human Services
DHS	Department of Homeland Security
DNI	Director of National Intelligence
DOC	Department of Commerce
DOD	Department of Defense
DOE	Department of Energy
DOEd	Department of Education
DOE/SC	DOE Office of Science
DOJ	Department of Justice
DOL	Department of Labor
DOS	Department of State
DOT	Department of Transportation
DOTreas	Department of the Treasury

DTRA	Defense Threat Reduction Agency (DOD)
EERE	[Office of] Energy Efficiency and Renewable Energy (DOE)
EHS	environment[al], health, and safety
ELSI	ethical, legal, and societal implications
ENM	engineered nanomaterial
EPA	Environmental Protection Agency
ERC	Engineering Research Centers (NSF)
ERDC	Engineer Research and Development Center
EU	European Union
FCRP	Focus Center Research Program
FDA	Food and Drug Administration (DHHS)
FHWA	Federal Highway Administration (DOT)
FPACE	Foundational Program to Advance Cell Efficiency
FPL	Forest Products Laboratory
FS	Forest Service (USDA)
HFCT	Hydrogen and Fuel Cells Technology program (DOE)
IAG	interagency agreement
IC	Intelligence Community
ISN	Institute for Soldier Nanotechnologies at MIT
ISO	International Organization for Standardization
LED	light-emitting diode
MEMS	microelectromechanical systems
MOU	memorandum of understanding
MRI	magnetic resonance imaging
MURI	Multidisciplinary University Research Initiative
nanoEHS	nanotechnology environment, health, and safety [research, etc.]
NASA	National Aeronautics and Space Administration
NCI	National Cancer Institute (DHHS/NIH)
NCL	Nanotechnology Characterization Laboratory (DHHS/NIH/NCI)
NEHI	Nanotechnology Environmental and Health Implications Working Group of the NSET Subcommittee
NERC	Nanosystems Engineering Research Centers (NSF)
NGO	nongovernmental organization
NIEHS	National Institute of Environmental Health Sciences (DHHS/NIH)

NIFA	National Institute of Food and Agriculture (USDA)
NIH	National Institutes of Health (DHHS)
NIOSH	National Institute for Occupational Safety and Health (DHHS/CDC)
NIST	National Institute of Standards and Technology (DOC)
NKI	Nanotechnology Knowledge Infrastructure (Nanotechnology Signature Initiative)
NNAP	National Nanotechnology Advisory Panel
NNCO	National Nanotechnology Coordination Office
NNI	National Nanotechnology Initiative
NRC	Nuclear Regulatory Commission
NRI	Nanoelectronics Research Initiative (NIST, NSF)
NRO	National Reconnaissance Office (IC/DNI)
NSEC	Nanoscale Science and Engineering Centers (NSF program)
NSET	Nanoscale Science, Engineering, and Technology Subcommittee of the NSTC
NSF	National Science Foundation
NSI	Nanotechnology Signature Initiative
NSRC	Nanoscale Science Research Centers (DOE)
NSTC	National Science and Technology Council
NTP	National Toxicology Program (DHHS/multiagency)
OECD	Organisation for Economic Co-operation and Development
OMB	Office of Management and Budget (Executive Office of the President)
ONR	Office of Naval Research

OSHA	Occupational Safety and Health Administration (DOL)
OSTP	Office of Science and Technology Policy (Executive Office of the President)
PCA	Program Component Area
PCAST	President's Council of Advisors on Science and Technology
PEN	Program(s) of Excellence in Nanotechnology (DHHS/NIH)
PPP	public–private partnership
PtD	Prevention through Design
PV	photovoltaic
QESST	Quantum Energy and Sustainable Solar Technologies (NSF/DOE Engineering Research Center)
RFA	request for applications
SBIR	Small Business Innovation Research Program
SETP	Solar Energy Technology Program (DOE/EERE)
STTR	Small Business Technology Transfer Research Program
TAPPI	Technical Association of the Pulp and Paper Industry
TONIC	Translation of Nanotechnology in Cancer
USDA	U.S. Department of Agriculture
USGS	U.S. Geological Survey
USITC	U.S. International Trade Commission
USPTO	U.S. Patent and Trademark Office (DOC)
WMD(s)	weapon(s) of mass destruction
WPN	Working Party on Nanotechnology (OECD)

APPENDIX C. CONTACT LIST

OSTP

Dr. Altaf H. Carim
Assistant Director for Nanotechnology
Office of Science and Technology Policy
Executive Office of the President
Washington, DC 20502
T: 202-456-7825
acarim@ostp.eop.gov

OMB

Ms. Celinda Marsh
Office of Management and Budget
Executive Office of the President
725 17th St., N.W.
Washington, DC 20503
T: 202-395-5835
Celinda_A._Marsh@omb.eop.gov

NNCO

Dr. Robert Pohanka
Director
National Nanotechnology
    Coordination Office
4201 Wilson Blvd.
Stafford II, Suite 405
Arlington, VA 22230
T: 703-292-8626; F: 703-292-9312
rpohanka@nnco.nano.gov

CPSC

Dr. Mary Ann Danello
Associate Executive Director
Directorate for Health Sciences
Consumer Product Safety Commission
4330 East-West Highway, Suite 600
Bethesda, MD 20814
T: 301-504-7237; F: 301-504-0079
mdanello@cpsc.gov

Dr. Treye Thomas
Leader, Chemical Hazards Program
Office of Hazard Identification & Reduction
Consumer Product Safety Commission
4330 East-West Highway,
    Suite 600
Bethesda, MD 20814
T: 301-504-7738; F: 301-987-2560
tthomas@cpsc.gov

DHS

Dr. Richard T. Lareau
Science and Technology Directorate
Department of Homeland Security
William J. Hughes Technical Center, Bldg. 315
Atlantic City International Airport
NJ 08405
T: 609-813-2760; F: 609-383-1973
Richard.Lareau@dhs.gov

Dr. Eric J. Houser
Department of Homeland Security
Office of Research and Development
Science and Technology Directorate
1120 Vermont Ave., N.W.
Washington, DC 20005
T: 202-254-5366
eric.houser@dhs.gov

DNI

Mr. Richard D. Ridgley
Chief Scientist, Advanced Systems and
    Technology Directorate
National Reconnaissance Office
14675 Lee Road
Chantilly, VA 20151
T: 703 808-3115; F: 703 808-2646
ridgleyr@nro.mil

Dr. Paul Kladitis
Branch Manager for Nanotechnology
Advanced Systems and Technology
    Directorate, Advanced Technology
    Office
14675 Lee Road
Chantilly, VA 20151
T: 703 808-4110, F: 703 808-2646
kladitis@nro.mil

DOC/BIS

Ms. Kelly Gardner
Export Policy Advisor
U.S. Department of Commerce
Bureau of Industry and Security
Office of National Security and Technology
    Transfer Controls, Rm. 2600
14th St. & Constitution Ave., N.W.
Washington, DC 20230
T: 202-482-0102; F: 202-482-3345
Kelly.Gardner@bis.doc.gov

DOD

Dr. Lewis Sloter
Associate Director, Materials & Structures
Office of the Assistant Secretary of Defense
    for Research and Engineering
AT&L (ASDR&E)
4800 Mark Center Dr.
Alexandria VA 22350-3600
T: 571-372-6511; F: 571-372-6500
Lewis.Sloter@osd.mil

Dr. Jeffrey DePriest
New Technologies Program Manager
Counterforce Systems Division
Defense Threat Reduction Agency
8725 John J. Kingman Road, MSC 6201
Fort Belvoir, VA 22060-6201
T: 703-767-4199; F: 703-767-5535
Jeffrey.DePriest@dtra.mil

Dr. Akbar S. Khan
Senior Microbiologist & S&T Manager
Basic & Supporting Research Div.
Defense Threat Reduction Agency
Fort Belvoir, VA 22060-6201
T: 703-767-3391; F: 703-767-1893
Akbar.Khan@dtra.mil

Dr. Gernot S. Pomrenke
Program Manager - Optoelectronics, THz
    and Nanotechnology
AFOSR/NE
Directorate of Physics and Electronics
Air Force Office of Scientific Research
875 North Randolph Street
Suite 325, Rm. 3112
Arlington, VA 22203-1768
T: 703-696-8426; F: 703-696-8481
gernot.pomrenke@afosr.af.mil

Dr. David M. Stepp
Chief, Materials Science Division
Army Research Office
AMSRD-ROE-M (Materials Science Division)
P.O. Box 12211
Research Triangle Park, NC 27709
T: 919-549-4329; F: 919-549-4399
david.m.stepp.civ@mail.mil

Dr. Charles R. Welch
Program Manager - Carbon Nanotube
    Technologies
Emeritus Director - Shock and Vibration
    Information Analysis Center
Information Technology Laboratory
U.S. Army Engineer R&D Center
3909 Halls Ferry Rd.
Vicksburg, MS 39180
T: 601.634.3297
Charles.R.Welch@usace.army.mil

DOE

Dr. Harriet Kung
Director, Office of Basic Energy Sciences
Office of Science
U.S. Department of Energy
SC-22/GTN
1000 Independence Ave., S.W.
Washington, DC 20585
T: 301-903-3081; F: 301-903-6594
Harriet.Kung@science.doe.gov

Dr. John C. Miller
Division of Chemical Sciences, Geosciences,
    and Biosciences
Office of Basic Energy Sciences
U.S. Department of Energy SC-
22.1/GTN
1000 Independence Ave., S.W.
Washington, DC 20585
T: 301-903-5806; F: 301-903-4110
John.Miller@science.doe.gov

Dr. Andrew R. Schwartz
Division of Materials Sciences and
    Engineering
Office of Basic Energy Sciences
U.S. Department of Energy SC-
22.2/GTN
1000 Independence Avenue SW
Washington, DC 20585-1290
T: (301) 903-3535; F: (301) 903-9513
andrew.schwartz@science.doe.gov

Dr. Brian Valentine
Advanced Manufacturing Office
U.S. Department of Energy
Headquarters Bldg., EE-2F
1000 Independence Avenue, S.W.
Washington, D.C. 20585-0121
T: 202-586-9741; F: 202-586-9234
Brian.Valentine@ee.doe.gov

DOEd

Mr. Peirce Hammond
Senior Advisor for Special Initiatives
U.S. Department of Education
Office of Vocational and Adult Education
Room 11-106 Potomac Center Plaza
400 Maryland Avenue, SW
Washington, DC 20202-7100
T: 202-245-6056
Peirce.Hammond@ed.gov

Dr. Krishan Mathur
U.S. Department of Education
Office of Postsecondary Education
1990 K Street, N.W., #6144
Washington, DC 20006-8544
T: 202-502-7512; F: 202-502-7877
krish.mathur@ed.gov

DOJ

Mr. Joseph Heaps
Deputy Director, In-situ Thermal
    Desorption Technology
Department of Justice
National Institute of Justice
Office of Science and Technology
810 7th Street, NW, Rm. 7206
Washington, DC 20531
T: 202-305-1554; F: 202-305-9091
Joseph.Heaps@usdoj.gov

DOL/OSHA

Dr. Janet Carter Senior
Health Scientist
Department of Labor
Occupational Safety and Health
    Administration
Room N3718
200 Constitution Ave., NW
Washington, DC 20210
T: 513-684-6030; F: 513-684-2630
Carter.Janet@dol.gov

DOS

Mr. Kenneth Hodgkins
Director
Office of Space & Advanced Technology
U.S. Department of State
OES/SAT, SA-23
1990 K Street, NW, Suite 410 Washington,
DC 20006
T: 202-663-2398; F: 202-663-2402
HodgkinsKD@state.gov

Dr. Chris Cannizzaro
Physical Science Officer
Office of Space & Advanced Technology
U.S. Department of State
OES/SAT, SA-23
1990 K Street, NW, Suite 410 Washington,
DC 20006
T: 202-663-2390; F: 202-663-2402
CannizzaroCM@state.gov

DOT

Mr. Peter Chipman
Senior Transportation Specialist
Research and Innovative Technology
    Administration
Department of Transportation
1200 New Jersey Ave., S.E.
Washington, DC 20590
T: 202-366-1637
Peter.Chipman@dot.gov

Dr. Jonathan Porter
Chief Scientist
Office of Research, Development, and
    Technology
Federal Highway Administration
U.S. Department of Transportation
Turner-Fairbank Highway Research Center
6300 Georgetown Pike
McLean VA 22101-2200
T: 202-493-3038; F: 202-493-3417
jonathan.porter@fhwa.dot.gov

DOTreas

Dr. John F. Bobalek
Program Manager
Office of Research and Technical Support
Bureau of Engraving and Printing
Rm. 201-29A
14th and C Streets, SW
Washington, DC 20228-0001
T: 202-874-2109; F: 202-927-7415
John.Bobalek@bep.treas.gov

EDA

Mr. Nishith Acharya
Director, Office of Innovation &
    Entrepreneurship & Senior Advisor to
    the Secretary of Commerce
U.S. Department of Commerce
1401 Constitution Avenue, NW
Suite 71014
Washington, DC 20230
T: 202-482-4068, F: 202-273-4781
nacharya@doc.gov

Mr. Thomas Guevara
Deputy Assistant Secretary for Regional
    Affairs
U.S. Department of Commerce
1401 Constitution Avenue, NW
Suite 71014
Washington, DC 20230
T: 202-482-5891, F: 202-273-4781
tguevara@eda.doc.gov

EPA

Dr. Tina Bahadori
National Program Director
Chemical Safety for Sustainability
Office of Research and Development
Mail Code 8101R
Environmental Protection Agency
1200 Pennsylvania Avenue, NW
Washington, DC 20460
T: 202-564-0428
Bahadori.Tina@epa.gov

Dr. Thabet Tolaymat
Interim Associate National Program
    Director
Office of Research and Development
Environmental Protection Agency
26 West Martin Luther King Drive
Cincinnati, OH 45268
T: 513-487-2860; F: 513-569-7879
Tolaymat.Thabet@epamail.epa.gov

Dr. Nora F. Savage
National Center for Environmental
    Research
Office of Research and Development
Mail Code 8722F
Environmental Protection Agency
1200 Pennsylvania Avenue, NW
Washington, DC 20460
T: 202-347-8104; F: 202-347-8142
Savage.Nora@epamail.epa.gov

Dr. Philip Sayre
Associate Director
Risk Assessment Division
Office of Pollution Prevention and Toxics
Environmental Protection Agency
1200 Pennsylvania Ave., NW
Mail Code 7403M
Washington, DC 20460
T: 202-564-7673; F: 202-564-7450
sayre.phil@epa.gov

FDA

Dr. Carlos Peña
Director, Emerging Technology Programs
Office of the Commissioner
FDA
10903 New Hampshire Avenue
WO Bldg 32 Room 4264
Silver Spring, MD 20993
T: 301-796-8521; F: 301-847-8617
carlos.pena@fda.hhs.gov

NASA

Dr. Michael A. Meador
Nanotechnology Project Manager, Game
    Changing Technologies
National Aeronautics and Space
    Administration
Glenn Research Center
21000 Brookpark Road
Cleveland, OH 44135
T: 216-433-9518; F: 216-977-7132
Michael.A.Meador@nasa.gov

Dr. Minoo N. Dastoor
Office of the Chief Technologist
National Aeronautics and Space
    Administration
NASA Headquarters, Rm. 5Q29
Washington DC 20546-0001
T: 202-358-4518; F: 202-358-3223
minoo.n.dastoor@nasa.gov

Dr. Meyya Meyyappan
National Aeronautics and Space
    Administration
Ames Research Center
MS 229-3
Moffett Field, CA 94035
T: 650-604-2616; F: 650-604-5244
m.meyyappan@nasa.gov

NIH

Dr. Lori Henderson
Program Director
Clinical Trials Branch/Cancer Imaging
    Program
Division of Cancer Treatment and
    Diagnosis
National Cancer Institute
6130 Executive Blvd.
Rockville, MD 20892
T: 301-451-6449; F: 301-480-3507
hendersonlori@mail.nih.gov

NIH/OD

Dr. Nancy E. Miller
Senior Science Policy Analyst
Office of Science Policy & Planning
Office of the Director
National Institutes of Health
1 Center Drive
Bldg 1, Rm. 218
Bethesda, MD 20892
T: 301-594-7742; F: 301-402-0280
Nancy.Miller1@nih.gov

NIH/FIC

Mr. George Herrfurth
Multilateral Affairs Coordinator
Division of International Relations
Fogarty International Center
National Institutes of Health
Building 31, Rm. B2C11
Bethesda, MD 20892-2220
T: 301-402-1615; F: 301-480-3414
herrfurg@mail.nih.gov

NIH/NCI

Dr. Piotr Grodzinski
Program Director, Nanotechnology in Cancer
Office of Technology & Industrial Relations
Office of the Director
National Cancer Institute
National Institutes of Health
31 Center Drive, 10A52
Bethesda, MD. 20892
T: 301-496-1550; F: 301-496-7807
grodzinp@mail.nih.gov

Dr. Scott McNeil
Director, Nanotechnology Characterization
    Laboratory
National Cancer Institute at Frederick
National Institutes of Health
P.O. Box B, Building 469
1050 Boyles Street
Frederick, MD 21702-1201
T: 301-846-6939; F: 301-846-6399
mcneils@ncifcrf.gov

NIH/NHLBI

Dr. Denis Buxton
Basic and Early Translational Research
    Program, DCVS
National Heart, Lung, and Blood Institute
National Institutes of Health
6701 Rockledge Drive, Rm. 8216
Bethesda, MD 20817
T: 301-435-0513; F: 301-480-1454
BuxtonD@nh bi.nih.gov

NIH/NIBIB

Dr. Steven Zullo
Gene and Drug Delivery, Nanotechnology
    Program Director
NIBIB, Division of Discovery Science and
    Technology
Democracy II-227
6707 Democracy Blvd
Bethesda, MD 20892
T: 301-451-4774; F: 301-480-1614
zullo@helix.nih.gov

NIH/NIEHS

Dr. John R. Bucher
Associate Director
National Toxicology Program
NIEHS
National Institutes of Health
P.O. Box 12233
Research Triangle Park, NC 27709
T: 919-541-4532; F: 919-541-4225
bucher@niehs.nih.gov

Dr. Christopher P. Weis
Toxicology Liaison/Senior Advisor
Office of the Director
NIEHS
National Institutes of Health
31 Center Drive, Room B1C02
Bethesda, MD 20892-2256
T: 301-496-3511
christopher.weis@nih.gov

NIH/NIGMS

Dr. Catherine Lewis
Director, Division of Cell Biology and
    Biophysics
National Institute of General Medical Sciences
National Institutes of Health
45 Center Drive, Rm. MSC 6200
Bethesda, MD 20892-6200
T: 301-594-0828; F: 301-480-2004
catherine.lewis@nih.hhs.gov

NIOSH

Dr. Charles L. Geraci, Jr.
Coordinator, Nanotechnology Research Center
National Institute for Occupational Safety
    and Health
4676 Columbia Parkway
Cincinnati, OH 45226
T: 513-533-8339
CGeraci@cdc.gov

Dr. Vladimir V. Murashov
Special Assistant to the Director
National Institute for Occupational Safety
    and Health
OD/NIOSH M/S-P12
395 E Street, S.W.
Suite 9200
Washington, DC 20201
T: 202-245-0668; F: 202-245-0628
vladimir.murashov@cdc.hhs.gov

NIST

Dr. Heather M. Evans
Senior Program Analyst
Program Coordination Office
National Institute of Standards and Technology
100 Bureau Drive, Stop 1060
Gaithersburg, MD 20899-1060
T: 301-975-4525; F: 301-216-0529
Heather.evans@nist.gov

Dr. Ajit Jillavenkatesa
Senior Standards Policy Adviser
Standards Coordination Office
National Institute of Standards and Technology
100 Bureau Drive, Stop 2100
Gaithersburg, MD 20899-2100
T: 301-975-8519; F: 301-975-4715
ajit.jilla@nist.gov

Dr. Debra L. Kaiser
Technical Program Director
Materials Measurement Science Division
National Institute of Standards and Technology
100 Bureau Drive, Stop 8520
Gaithersburg, MD 20899-8520
T: 301-975-6759; F: 301-975-5334
debra.kaiser@nist.gov

Dr. Lloyd J. Whitman
Deputy Director, Center for Nanoscale Science
    and Technology
National Institute of Standards and Technology
100 Bureau Drive, Stop 6200
Gaithersburg, MD 20899-6200
T: 301-975-8002; F: 301-975-8026
lloyd.whitman@nist.gov

NRC

Mr. Stuart Richards, Deputy Director
Division of Engineering
Office of Nuclear Regulatory Research
Nuclear Regulatory Commission
Washington, DC 20555-0001
T: 301-251-7616
Stuart.Richards@nrc.gov

NSF

Dr. Mihail C. Roco
Senior Advisor for Nanotechnology
National Science Foundation
Directorate for Engineering
4201 Wilson Blvd., Suite 505
Arlington, VA 22230
T: 703-292-8301; F: 703-292-9013
mroco@nsf.gov

Dr. Parag R Chitnis
Deputy Director, Directorate for Biological
   Sciences
National Science Foundation
4201 Wilson Blvd., Suite 605N
Arlington, VA 22230
T: 703-292-8440; F: 703-292-9061
pchitnis@nsf.gov

Dr. Thomas Rieker
Program Director, Division of Materials
   Research
National Science Foundation
Directorate for Mathematical and Physical
   Sciences
4201 Wilson Blvd., Suite 1065
Arlington, VA 22230
T: (703) 292-4914, F: (703) 292-9035
trieker@nsf.gov

Dr. Grace J. Wang
Program Manager, SBIR/STTR
Office of Industrial Innovation
National Science Foundation
4201 Wilson Blvd., Rm. 590
Arlington, VA 22230
T: 703-292-2214; F: 703-292-9057
jiwang@nsf.gov

Dr. Khershed Cooper
Program Director, Nanomanufacturing
National Science Foundation
Directorate for Engineering
4201 Wilson Blvd., Suite 505
Arlington, VA 22230
T: 703-292-8301; F: 703-292-9013
kcooper@nsf.gov

USDA/ARS

Dr. Robert Fireovid,
National Program Leader - Biobased
   Products and Biorefining
USDA Agricultural Research Service
5601 Sunnyside Avenue
GWCC 4-2264
Beltsville, MD 20705-5140
T: 301-504-4774; F: 301-504-6231
Robert.Fireovid@ars.usda.gov

USDA/NIFA

Dr. Hongda Chen
National Program Leader, Bioprocess
   Engineering/Nanotechnology National
Institute of Food and Agriculture U.S.
Department of Agriculture
1400 Independence Ave., SW
Mail Stop 2220
Washington, DC 20250-2220
T: 202-401-6497; F: 202-401-4888
HCHEN@nifa.usda.gov

USDA/FS

Dr. Theodore H. Wegner
Assistant Director
Forest Products Laboratory
USDA Forest Service
One Gifford Pinchot Drive
Madison, WI 53726-2398
T: 608-231-9434; F: 608-231-9567
twegner@fs.fed.us

Dr. World Nieh
National Program Leader, Forest Products
   and Utilization
USDA Forest Service
1400 Independence Ave., SW
Mailstop Code: 1114
Washington, DC 20250-1114
T: 703-605-4197; F: 703-605-5137
wnieh@fs.fed.us

USGS

Dr. Sarah Gerould
Staff Scientist, Ecosystems Mission Area
U.S. Geological Survey
Mail Stop 301 National Center
12201 Sunrise Valley Drive
Reston, VA 20192
T: 703-648-6895; F: 703-648-4238
sgerould@usgs.gov

USITC

Ms. Elizabeth R. Nesbitt
International Trade Analyst for Bio-
   technology and Nanotechnology
Chemicals and Textiles Division
Office of Industries
U.S. International Trade Commission
500 E Street, S.W.
Washington, DC 20436
T: 202-205-3355; F: 202-205-3161
elizabeth.nesbitt@usitc.gov

USPTO

Mr. David R. Gerk
Patent Attorney
Office of Policy and External Affairs
Patent and Trademark Office
P.O. Box 1450
Alexandria, VA 22313
T: 571-272-9300; F: 571-273-0085
david.gerk@uspto.gov

Mr. Bruce Kisliuk
Deputy Commissioner for Patent
   Administration
Patent and Trademark Office
P.O. Box 1450
Alexandria, VA 22313
T: 571-272-8800
Bruce.Kisliuk@uspto.gov

Ms. Gladys Corcoran
Technology Center Group Director
Patent and Trademark Office
P.O. Box 1450
Alexandria, VA 22313
T: 571-272-1300
Gladys.Corcoran@uspto.gov

**Additional Staff Contacts,
National Nanotechnology Coordination
Office (NNCO)**

Dr. Tarek R. Fadel
Policy Analyst
NSET and NEHI Executive Secretary
NNCO
4201 Wilson Blvd.
Stafford II, Suite 405
Arlington, VA 22230
T: 703-292-7926; F: 703-292-9312
tfadel@nnco.nano.gov

Dr. Lisa E. Friedersdorf
Sr. Policy Analyst, NSI Project Manager
NNCO
4201 Wilson Blvd.
Stafford II, Suite 405
Arlington, VA 22230
T: 703-292-4533; F: 703-292-9312
lfriedersdorf@nnco.nano.gov

Mr. Geoffrey M. Holdridge
Policy Analyst and Contract Staff Director
NNCO
4201 Wilson Blvd.
Stafford II, Suite 405
Arlington, VA 22230
T: 703-292-4532; F: 703-292-9312
gholdrid@nnco.nano.gov

Dr. Mike Kiley
Industry and State Liaison
NILI Executive Secretary
NNCO
4201 Wilson Blvd.
Stafford II, Suite 405
Arlington, VA 22230
T: 703-292-4399; F: 703-292-9312
mkiley@nnco.nano.gov

Ms. Marlowe Newman
Communications Director
NNCO
4201 Wilson Blvd.
Stafford II, Suite 405
Arlington, VA 22230
T: 703-292-7128; F: 703-292-9312
mnewman@nnco.nano.gov